44

derangements

and the shape of persistence

Thad Roberts

44 derangements and the shape of persistence

second edition

By Thad Roberts

First print date 03-17-2022

Published in the United States by Thad Roberts

ISBN 979-8-9859554-2-2

Cover by Jeff Chapple
Figures by Thad Roberts and Jeff Chapple
    Original hyperbolic figure eight knot STL by Henry Segerman
    https://www.thingiverse.com/thing:1668611

Other books by Thad:

    Einstein's Intuition: Visualizing Nature in Eleven Dimensions

    Moon Rock: Mare Crisium

    Passages

    A Perfect Universe

    Source Code: the balance of persistence

to persistence

Preface: the search

"The task is not to see what has never been seen before, but to think what has never been thought before about what you see every day."

Erwin Schrödinger

In an early attempt to construct a story that accounts for reality's persistent physical properties, Plato advanced the idea that the elemental building blocks of geometry are also the elemental building blocks of reality; envisioning the Platonic solids (the 5 elemental shapes constructible from faces with equal length sides) as *atoms* of fire, air, earth, water, and aether—the primordial container of it all.

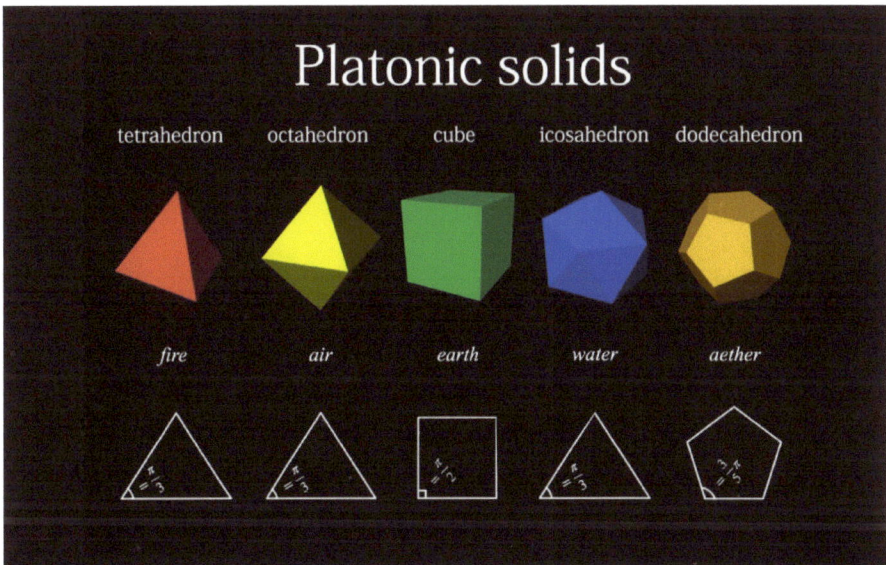

The 5 Platonic solids

The reach of Plato's *idea* triggered a categorical awakening. If *geometry* plays a role in the *construction of reality*, then it becomes possible to walk ourselves out of the cave of ignorance. A *geometry* is something we can say more about, something we can investigate, measure, and discover the internal logic of. In other words, geometry gives us a way of looking closer into things, a way of seeking greater clarity and finding it, a way of exercising the powers of self-transformation.

For the first couple of eons after Plato's insight lit the caverns of consciousness, the search for Nature's implicitly obtainable logic, or answering the call of existence and making one's best attempt to make sense of things, meant studying the Platonic solids. If you wanted to uncover the secrets of reality, you would spend your hours staring at these elemental shapes, trying to notice anything else they might have to say.

After 2000 years of attempts to see further into the Platonic solids and glean more from them, Leonard Euler came along and, with little more than a glance, noticed that they are all connected by a single elementary geometric relationship.

Every Platonic solid maintains the same balance of zero, one, and two-dimensional features (vertices, edges, faces). Each shape's number of vertices, minus its number of edges, plus its number of faces is always equal to 2.

$$\text{vertices} - \text{edges} + \text{faces} = 2$$

$$V - E + F = 2$$

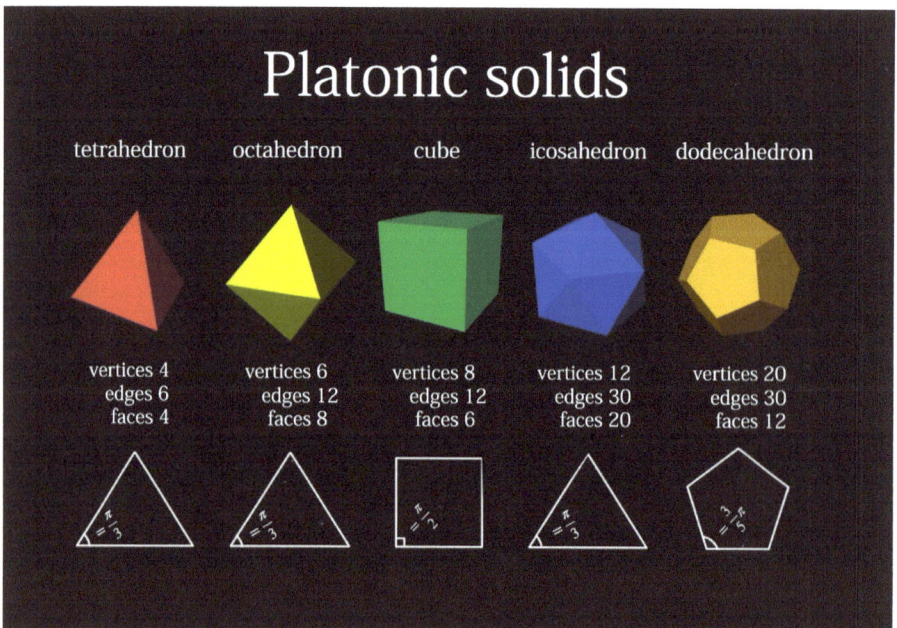

Platonic solids

tetrahedron	octahedron	cube	icosahedron	dodecahedron
vertices 4 edges 6 faces 4	vertices 6 edges 12 faces 8	vertices 8 edges 12 faces 6	vertices 12 edges 30 faces 20	vertices 20 edges 30 faces 12

Euler's surprise relationship. Until Euler came along the number of edges these objects had were "hidden features", missed by everyone.

Despite the intense interest in these geometric forms, and the generations of people specifically searching them for further clues, nobody had ever found what Euler found. Why? It doesn't take much time to count up the properties of these shapes and compare them. Yet nobody had. Why not?

The answer is that, before Euler, nobody had imagined that an object's *number of sides* was a feature worthy of attention. They had never seen anyone else care about such a feature, so it just hadn't occurred. Euler was the first to count up these features and compare them because he was the first to imagine the option. He woke us up to that concept (and *many* more), lifting the ceiling of what it means *to think* and, therefore, *to be*.

Conscious thinkers have two routes to new thought. Through example, or through genuine new reaches of imagination. Just as it is usually easier to verify an answer than to find one, it is usually easier to follow a line of logic laid out for you than to lay one out for yourself. Which is why conscious thinkers almost always follow.

The trouble is that, if we are always just following, then we can get seriously stuck. If nobody has ever shown us an example of being interested in the number of edges an object has, we can all stare at something for 2000 years without seeing what's right in front of our eyes.

Euler looked into the world with new eyes, under his *own* investigation, instead of looking only as he had been directed to. And, as a consequence, the world inherited new conceptual powers.

Another famous example of how difficult it can be to *look into reality* and *see it*, comes from our history of conceptualizing gravity. For millennia, people pondered *why* things fall to the ground, but it wasn't until Galileo came along that anyone thought to measure *how* things fall under the influence of gravity.

Galileo's measurements could be performed by nearly anyone within minutes, with no cost; like rolling a ball down an inclined plane and marking the ball's position at successive increments of time and then comparing the distance between those marks. But before Galileo nobody had tried. Nobody had *thought* to.

By extending the powers of *measurement* to gravity, Galileo became the first to see it more clearly; famously finding that gravity's action has an inverse square dependence on distance. In other words, he discovered that *gravity is geometric*.

This fact about gravity had always been true, it had always been staring us in the face, screaming from every freefall occurrence ever witnessed, but nobody had seen it before. Because nobody thought of *gravity* as a thing to physically measure, a thing to *look into* that way.

Through his efforts Galileo came to one of the greatest insights of all time, that the truth about reality is right in front of our eyes. The task is to think about it more clearly, to more explicitly *conceive* its true form.

> *"Philosophy is written in this grand book, the universe, which stands continually open to our gaze. But the book cannot be understood unless one first learns to comprehend the language and read the letters in which it is composed. It is written in the language of mathematics, and its characters are triangles, circles, and other geometric figures without which it is humanly impossible to understand a single word of it; without these, one wanders about in a dark labyrinth."*
>
> *Galileo Galilei*

Today's best geometric description of the fundamental structure of reality is called *quantum field theory*. Correctly parametrized, this construction produces all of the actions of quantum mechanics *and* special relativity, a feat that earns it the title of being the pinnacle achievement of science.

The problem is that, although quantum field theory makes the right predictions, nobody understands why it has that particular construction. That is, the parameters of quantum field theory have no story themselves. Every single one of them remains utterly unexplained, known from experimental measurement only.

In short, we cannot explain *why* quantum field theory is constructed as it is, we cannot predict its parameters to any degree of accuracy at all, but once we construct our field theory with those particular parameters the actions of quantum mechanics and special relativity are reproduced in full.

> *"Look into Nature, and then you will understand it better."*
> *Albert Einstein*

Today, any serious investigation of reality means staring at the parameters of quantum field theory, in search of the logic connecting them. That is, the modern version of the quest to tell the accurate story

10

of reality's construction (now called *the theory of everything*) literally boils down to explaining the parameters of quantum field theory, explaining why they are what they are.

the parameters of quantum field theory

$m_e = 9.1093837015(28) \times 10^{-31} \ kg$	electron mass
$m_+ = 1.67262192369(51) \times 10^{-27} \ kg$	proton mass
$m_N = 1.67492749804(95) \times 10^{-27} \ kg$	neutron mass
$m_\mu = 1.883531627(42) \times 10^{-28} \ kg$	muon mass
$m_Z = 1.625566(38) \times 10^{-25} \ kg$	Z boson mass
$m_W = 1.43288(21) \times 10^{-25} \ kg$	W boson mass
$m_\tau = 3.16754(21) \times 10^{-27} \ kg$	tau mass
$m_H = 2.2315(28) \times 10^{-25} \ kg$	Higgs boson mass
$m_t = 3.084(07) \times 10^{-25} \ kg$	truth (top) quark mass
$m_c = 2.272(63) \times 10^{-27} \ kg$	charm quark mass
$m_b = 7.45(07) \times 10^{-27} \ kg$	beauty (bottom) quark mass
$m_s = 1.69(16) \times 10^{-28} \ kg$	strange quark mass
$m_u = 3.92(89) \times 10^{-30} \ kg$	up quark mass
$m_d = 8.3(07) \times 10^{-30} \ kg$	down quark mass
$m_{\nu_\tau} = > 0 \ kg \ ???$	tau neutrino mass
$m_{\nu_\mu} = > 0 \ kg \ ???$	muon neutrino mass
$m_{\nu_e} = > 0 \ kg \ ???$	electron neutrino mass

$\alpha = 7.2973525698(24) \times 10^{-3}$	fine-structure constant
$e = 1.602176565(35) \times 10^{-19} \ C$	electron charge
$\alpha_G = 1.7518(21) \times 10^{-45}$	gravitational coupling constant
$\mu_0 = 1.256637061 \ldots \times 10^{-6} \ mkg/C^2$	magnetic constant
$S_{mi} = 4.419 \times 10^9 \ kg/sC$	Schwinger magnetic induction
$R_K = 2.58128074434(84) \times 10^4 \ m^2kg/sC^2$	von Klitzing constant
$H_C = 3.87404614(17) \times 10^{-5} \ C^2/m^2kg$	quantized Hall conductance
$\Phi_0 = 2.067833848 \ldots \times 10^{-15} \ m^2kg/sC$	magnetic flux constant
$K_J = 4.835978484 \times 10^{14} \ sC/m^2kg$	Josephson constant
$q_c = 3.6369475516(11) \times 10^{-4} \ m^2/s$	quantum of circulation
$G_0 = 7.748091729 \ldots \times 10^{-5} \ sC^2/m^2kg$	conductance quantum
$c_2 = 1.438776877 \ldots \times 10^{-2} \ m \ K$	2nd radiation constant
$\hbar = 1.054571726(47) \times 10^{-34} \ m^2kg/s$	Planck's constant
$\kappa = 8.9875517923(14) \times 10^9 \ m^3kg/s^2C^2$	Coulomb's constant
$\varepsilon_0 = 8.8541878128(13) \times 10^{-12} \ s^2C^2/m^3kg$	electric constant
$c = 2.99792458 \times 10^8 \ m/s$	speed of light

$F = 9.648533212 \ldots \times 10^4 \ C/mol$ — Faraday constant
$g_N = -3.82608545(90)$ — neutron g-factor
$G = 6.67384(80) \times 10^{-11} \ m^3/s^2 kg$ — gravitational constant
$k_B = 1.380649 \times 10^{-23} \ m^2 kg/s^2 K$ — Boltzmann constant
$\omega_c = 7.763441 \times 10^{20} \ 1/s$ — Compton angular frequency
$c_{1L} = 1.191042869(53) \times 10^{-16} \ m^4 kg/s^3$ — spectral radiance
$c_1 = 3.741771852 \ldots \times 10^{-16} \ m^4 kg/s^3$ — 1st radiation constant
$Z_0 = 3.76730313668(57) \times 10^2 \ m^2 kg/sC^2$ — characteristic impedance
$a_0 = 5.2917721092(17) \times 10^{-11} \ m$ — Bohr electron radius
$r_e = 2.8179403227(19) \times 10^{-15} \ m$ — classical electron radius
$E_h = 4.3597447222071(85) \times 10^{-18} \ m^2 kg/s^2$ — Hartree energy
$\lambda_C = 2.4263102389(16) \times 10^{-12} \ m$ — Compton wavelength
$\mu_B = 9.274009994(57) \times 10^{-24} \ m^2 C/s$ — Bohr magneton
$\mu_N = 5.050783699(31) \times 10^{-27} \ m^2 C/s$ — Nuclear magneton
$N_A = 6.02214076 \times 10^{23} \ 1/mol$ — Avogadro constant
$\sigma = 5.670374419 \times 10^{-8} \ kg/s^3 K^4$ — Stefan-Boltzmann constant
$R = 8.314462618 \ m^2 kg/s^2 K \ mol$ — molar gas constant
$\gamma_+ = 2.6752218744(11) \times 10^8 \ s/kg \ C$ — proton gyromagnetic ratio
$m_u = 1.66053906660(50) \times 10^{-27} \ kg$ — atomic mass constant
$\sigma_e = 6.6524616(18) \times 10^{-29} \ m^2$ — electron Thomson cross section
$R_\infty = 1.0973731568539(55) \times 10^7 \ 1/m$ — Rydberg constant
$g_\mu = -2.00233184122(82)$ — muon g-factor
$g_e = -2.00231930436256(35)$ — electron g-factor
$g_+ = +5.5856946893(16)$ — proton g-factor
$N_\mu = -9.6623647(23) \times 10^{-27} \ m^2 C/s$ — neutron magnetic moment
$b_{entropy} = 3.002916077 \times 10^{-3} \ mK$ — Wien entropy constant
$r_N = 8.0(10) \times 10^{-16} \ m$ — neutron radius
$r_+ = 8.414(19) \times 10^{-16} \ m$ — proton radius

Where the digits in the parentheses define the measurement error in the preceding two digits (e.g. 8.414(19) means 8.414 ± 0.019), and the neutrino masses are only known to be non-zero.

This book unveils the story of these numbers, discovering that all of the parameters of quantum field theory (and general relativity) are mapped by the combinatorial logic of the minimal self-balanced arena. That is, in this book we notice for the first time that the minimally partitioned arena, internally defined by the minimal possible volume complement (the hyperbolic figure eight knot) externally counterbalanced as the n-hypersphere of maximal volume, is responsible for setting the stage of physical reality—and for giving it *all* of its constructive parameters.

The layout of this book is as follows.

In Chapters 0-4 we decompose the minimal arena into its 5 unique balances.

In Chapter 5 we discover that the boundary conditions of this decomposition precisely predict the Planck constants, the limiting boundaries of the 5 fundamental bases or *dimensions of measure* in physics (time, space, charge, mass, and temperature).

In Chapters 6-8 we observe that the external facing boundaries of the minimal arena connect under hyperbolic vortex arrangement, and this connection precisely partitions the Planck charge and Planck mass boundaries into the charge and mass assignments that define the fundamental particles of matter.

In Chapter 9 we discover that (under universal binomial factorization) the 44 derangements of this 5-dimensional minimally partitioned arena define the constants of Nature.

In Chapters 10-11 we notice that the Planck boundaries maintain a union that ideally hyperbolically connects (defined by *e*, the base of hyperbolic logarithms), *and* ideally hyperbolically partitions (defined by the gamma function).

And in Chapters 12-13 we discover that the Riemann zeta function, the octonions, and the logic of calculus map the external combinatorial features of the minimal arena.

In summary, in this book we discover that the union of the minimum volume complement (the hyperbolic figure eight knot) and its external counterpart (the *n*-hypersphere of maximal volume) defines the *minimal arena of persistence*. The division parameters of that minimal arena set the fundamental bases of physics (time, space, charge, mass, and temperature). The external charge and mass boundaries of that arena partition into the charge and mass assignments that define the fundamental particles of matter. And the 44 unique derangements of this 5-dimensional arena define the constants of Nature.

The internal division parameters of the simplest possible arena (the minimal self-closed partitioned geometry) elegantly set the constructive parameters of reality.

Chapter 0: the simplest manifold

In the mid 1970's Robert Riley and Troels Jørgensen independently discovered that the figure eight knot emits a complement with hyperbolic structure. This was the first known example of a hyperbolic knot—a closed boundary formed under hyperbolic balance.

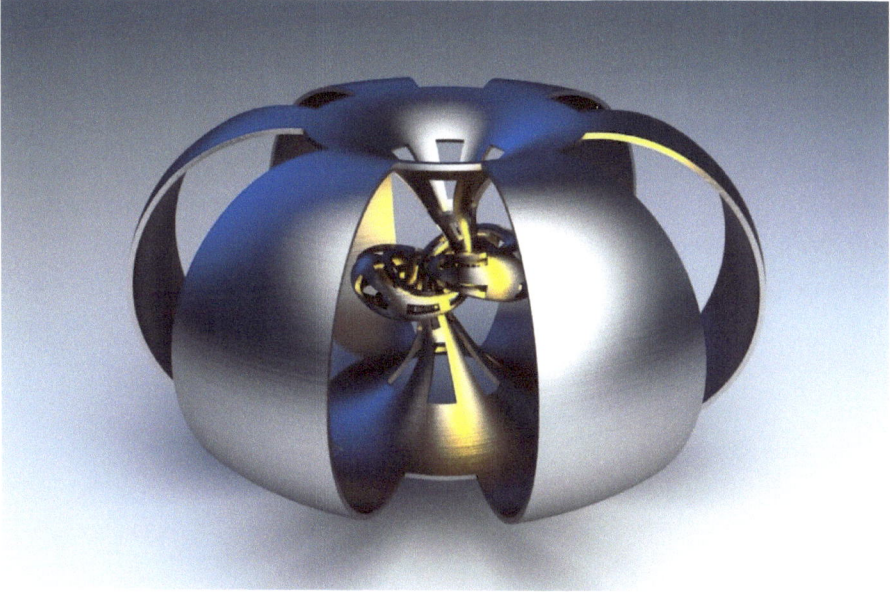

The complement of the hyperbolic figure-eight knot (portrayed here with cut-outs to allow visibility) is the finite volume bounded by the simple closed geodesics that trace out the figure-eight knot.

The hyperbolic figure eight knot is a double cover of the Gieseking manifold, the simplest among all non-compact hyperbolic 3-manifolds.[1] Its internal volume complement, equal to twice Gieseking's constant (G_{Gi}), defines the smallest possible self-closed volume complement (V_{fe}).

---

[1] A manifold is a geometry that self-closes, one whose boundary has been removed. For example, to transform a rectangle (a 2-dimesnional geometric form) into a manifold we fold it around on itself one way to form a cylinder, removing 2 of the boundaries by connecting them, then we twist that tube into a closed torus by connecting the circular ends of the cylinder, removing the remaining boundaries.

$$V_{fe} = 2G_{Gi}$$

$G_{Gi} = 1.01494160640965 \ldots$ Gieseking's constant
$V_{fe} = 2.02988321281930 \ldots$ figure eight knot complement volume

Achieving this minimum manifold means separating a finite domain into 2 regions of action; a volume whirling about *inside* the minimum complement division boundary, the hyperbolic figure eight knot, and a volume *outside* that boundary, inversely partitioning as the *n*-hypersphere of maximal volume.

Together, these internal and external partition actions define the minimal persistent stage (universe), maintaining the simplest possible perpetual balance of boundary conditions. That is to say, *persistence* is *minimally geometrically bound* by the partition structure of this 5-dimensional hyperbolically balanced arena.

$$V_{fe} = i\left[-Li_2\left((-1)^{\frac{1}{3}}\right) + Li_2\left(-(-1)^{\frac{2}{3}}\right)\right]$$

$$-Li_2\left((-1)^{\frac{1}{3}}\right) = -\left(\frac{4\pi}{\Gamma(n)}\right)^2 - G_{Gi}\, i$$

$$Li_2\left(-(-1)^{\frac{2}{3}}\right) = \left(\frac{4\pi}{\Gamma(n)}\right)^2 - G_{Gi}\, i$$

Where π = Archimedes' constant, G_{Gi} = Gieseking's constant, V_{fe} = the complement volume of the figure eight knot, Li_2 is the hyperbolic dilogarithm (the polylogarithm of order 2), $\Gamma(s)$ = the gamma function which encodes hyperbolically balanced partitions, $n = 5$, and $i = \sqrt{-1}$ is the imaginary unit.

When William Thurston looked into the internal part of this ideal geometry (the hyperbolic figure eight knot) he discovered that it decomposes into a union of 2 regular ideal hyperbolic tetrahedra. He then famously formulated his geometrization conjecture, claiming that all 3-manifolds admit a geometric decomposition involving 8 geometries, most of which are hyperbolic. Grigori Perelman proved this conjecture in 2002-2003.

To completely characterize this ideal geometry (internally and externally) we will now fully decompose it into its 5 elementary partition balances.[2]

---

[2] The internal and external expressions of this balance are maintained under Möbius connection. That is, the same circles connect into 2 different (internal and external) constructive geometries. In the next 2 figures, I have connected the circles in both ways (into their internal and external expressions) and placed them side by side, to make the internal/external arrangements easier to compare. The geometry on the left is portrays the hyperbolic figure eight knot, while the geometry on the right portrays the n-hypersphere of maximal volume).

Chapter 1: the minimum partition balance

The action of the minimal arena is 5-dimensional, it decomposes into 5 unique geometric balances (balance 0-4).

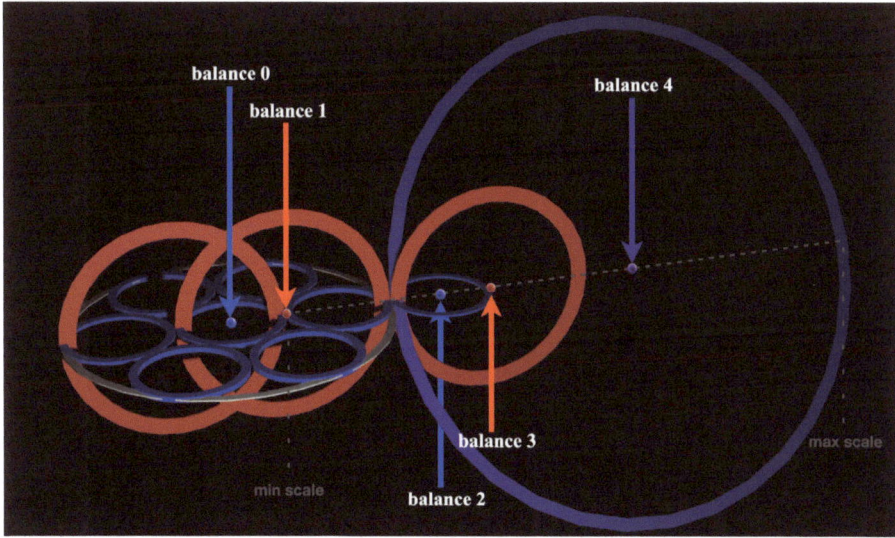

The partition balance of the minimal arena balances 2 internal actions (balance 0-1) against 3 unique external actions (balance 2-4).

general decomposition map of the minimal arena

$(\text{inner action})_4{}^{-1}(\text{outer rotation})_4{}^{-1} = d_4{}^{-1}$ $(\text{boundary 4})^{-1}$

$(\text{the external reference}) = 1$ balance 4

$(\text{inner action})_3(\text{outer rotation})_3 = d_3$ balance 3

$(\text{inner action})_2(\text{outer rotation})_2 = d_2$ balance 2

$(\text{inner action})_4(\text{outer rotation})_4 = d_4$ boundary 4

$(\text{inner action})_1(\text{outer rotation})_1 = d_1$ balance 1

$(\text{inner action})_0(\text{outer rotation})_0 = d_0$ balance 0

Where the union of balance 0 and balance 1 defines the internal double-cover domain (the hyperbolic figure eight knot), balance 4 (created by the union of balance 2 and balance 3) defines the projective balance of external domain (the n-hypersphere of maximal volume), and $d_k =$ the number of derangements involved in the kth balance.

The most external balance (balance 4) = 1. That is, the projective balance point of this minimally decomposed arena defines the minimum unitary self-divisible reference. (Things that divide 1 define dimensions.)

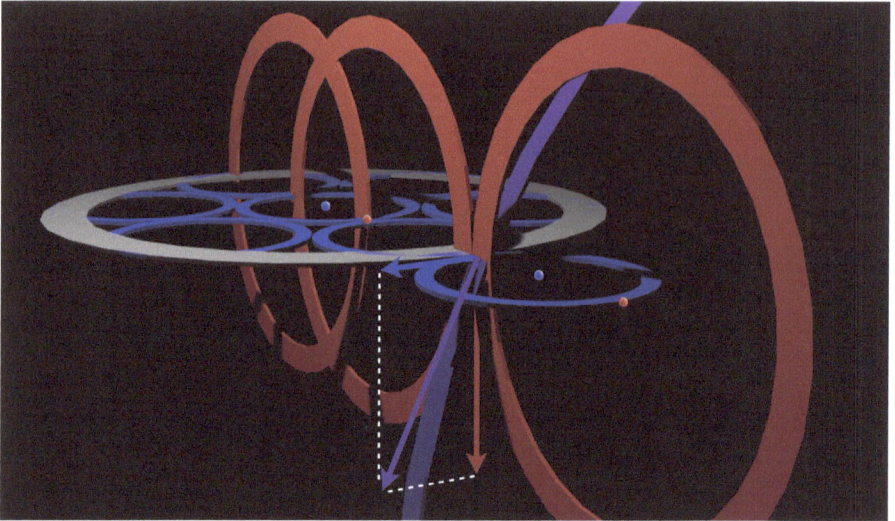

The combined action of balance 2 and 3 (blue and red arrows) rotates the external domain under balance 4 (purple).

The derangements of this balance close under ideal split factorization, internally defined as the square root of their difference $\left(\sqrt{d_0 - d_1} \right)$ and externally defined as *doubled* and *squared* symmetric expressions of those internal factorizations $\left(2\left(\sqrt{d_0 - d_1} + k \right)^2 \right)$, closed under simple quadratic connection ($+k$).

derangement balance

external derangements

$$d_3 = 2\left(\sqrt{d_0 - d_1} - 1 \right)^2 = 2(3 - 1)^2 = 8$$
$$d_2 = 2\left(\sqrt{d_0 - d_1} \pm 0 \right)^2 = 2(3 \pm 0)^2 = 18$$
$$d_4 = 2\left(\sqrt{d_0 - d_1} + 1 \right)^2 = 2(3 + 1)^2 = 32$$

internal derangements

$$d_1 = \boldsymbol{bn} = 35$$
$$d_0 = \boldsymbol{!n} = 44$$

Where $n = 5$ the number of unique rotations partitioning the minimal arena, $!n =$ the derangement function, $!n = 44$ the number of derangements available to 5 rotations, $b = 7$ the break in scale symmetry between the 2 internal balances (see Chapter 3), and the 3 external phases $k = \{-1,0,1\}$ define the 3 surfaces of hyperbolic geometry encoded by the quadratic form:

$$x^2 + y^2 - z^2 = k$$

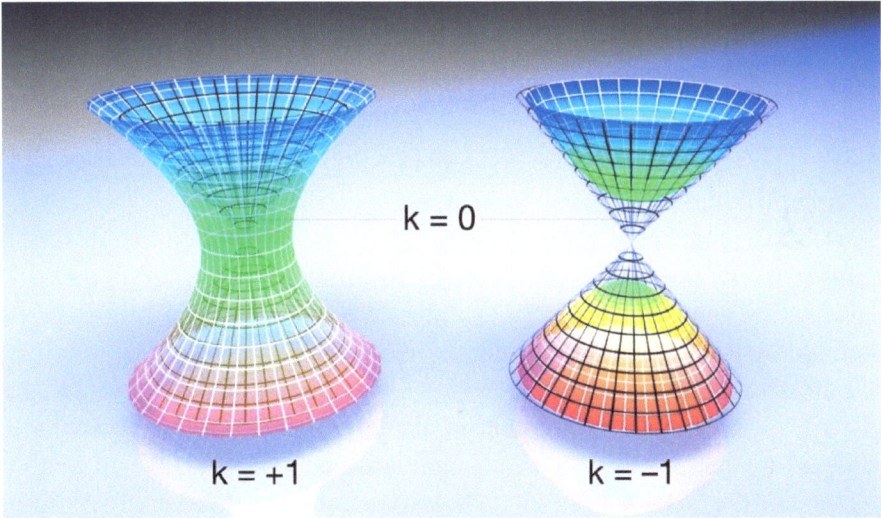

The 3 external surfaces of hyperbolic geometry: $k = +1$ defines the one-sheeted hyperploid, $k = 0$ defines the null cone (the conic division boundary of the external domain), and $k = -1$ defines the two-sheeted hyperploid.

    This minimum symmetry constraint (being maintained under finite rotation via split-symmetric division, under closed external quadratic connection) allows us to completely specify the right-hand side of the minimal arena's decomposition map (its balance of derangements).

    To sharpen this map further, we note that no matter how complex the interior action of a particular balance is, its exterior expression will always be a hyperbolic rotation of some magnitude. In other words, every (outer rotation)$_k = e^{\Phi k}$, where $e =$ Euler's number.

    This updates our decomposition map to:

decomposition map of the minimal arena

$$(\text{inner action})_4{}^{-1} e^{-\phi_4} = d_4{}^{-1} \qquad (\text{boundary 4})^{-1}$$

$$(\text{external reference}) = 1 \qquad\qquad \text{balance 4}$$

$$(\text{inner action})_3 \, e^{\phi_3} = d_3 \qquad\qquad \text{balance 3}$$

$$(\text{inner action})_2 \, e^{\phi_2} = d_2 \qquad\qquad \text{balance 2}$$

$$(\text{inner action})_4 \, e^{\phi_4} = d_4 \qquad\qquad \text{boundary 4}$$

$$(\text{inner action})_1 \, e^{\phi_1} = d_1 \qquad\qquad \text{balance 1}$$

$$(\text{inner action})_0 \, e^{\phi_0} = d_0 \qquad\qquad \text{balance 0}$$

Where ϕ_0, ϕ_1, ϕ_2, ϕ_3, & ϕ_4 are the rotation magnitudes for each balance, $d_0 = 44$ the number of derangements under the 0th rotation, $d_1 = 35$ the number of derangements under the 1st rotation, $d_2 = 18$ the number of derangements under the 2nd rotation, $d_3 = 8$ the number of derangements under the 3rd rotation, $d_4 = 32$ the number of derangements under the 4th rotation, $e =$ Euler's number, the union of balance 0 and balance 1 defines the internal double-cover domain (the internal complement of the hyperbolic figure eight knot), and balance 4 (created by the union of balance 2 and balance 3) defines the projective balance of the external domain (the n-hypersphere of maximal volume).

To resolve the values of those rotation magnitudes we need to specify the minimal arena's balance of inner actions. Let's begin by specifying its minimum cyclic expression (balance 0).

Chapter 2: the minimum limit of persistence

Let (balance 0) define the complete derangement ($d_0 = !n$) of 5 unique rotations.[3] And let the geometry of this minimum cyclic action be defined as a circularly closed action πr^2 trivially balanced against an external counter-rotation e^{ϕ_0}.

$$\pi r^2 e^{\phi_0} = !n \qquad\qquad \text{balance 0}$$

Where $\pi =$ Archimedes' constant, $e =$ Euler's number, $n = 5$ the number of unique rotations partitioning the minimal arena, $d_0 = !n =$ 44 the number of derangements available to those 5 rotations, and $\phi_0 =$ the rotation magnitude of this minimum cyclic action.

To represent this minimum cycle being maintained under trivial hyperbolic construction (factoring into rotations that counterbalance under ideal complex division—splitting into equal but opposite internal rotations, each absorbing half the input) we set the circle's radius (r) equal to the hyperbolic sine function.

$$\frac{1}{2}(e^x - e^{-x}) = sinh(x) = r$$

And we set its argument (x) equal to the minimal representation of square split division—a number that constructively possesses complex four-fold symmetry.

$$x = \left(\frac{1}{2}\right)^2 = \left(-\frac{1}{2}\right)^2 = -\left(\frac{1}{2}i\right)^2 = -\left(-\frac{1}{2}i\right)^2$$

$$\frac{1}{2}\left(e^{\left(\frac{1}{2}\right)^2} - e^{-\left(\frac{1}{2}\right)^2}\right) = sinh\left(\left(\frac{1}{2}\right)^2\right) = r$$

Plugging this ideally hyperbolic r into our equation, we arrive at a precise characterization of the minimum cyclic limit of *persistence*—an equation for the smallest possible cyclic partition action.

――――――――――――――――――

[3] A derangement is a combinatorial permutation that has no fixed points. That is, it is a balanced rearrangement whose parts all play active roles.

$$\pi\left(sinh\left(\left(\frac{1}{2}\right)^2\right)\right)^2 e^{\phi_0} = !\,n \qquad\qquad \text{balance 0}$$

Where $\phi_0 = 5.39125836832313\ldots + 2\pi k\,i$ the magnitude of the external rotation of balance 0, $\pi =$ Archimedes' constant, $sinh(x) =$ the hyperbolic sine function, $n = 5$ the number of unique rotations partitioning the minimal arena, and $d_0 = !\,n = 44$ the number of derangements available to 5 rotations.

Since this equation defines the minimum cyclic partition action within the minimum self-balanced arena, it holds the honor of defining the *ultimate* boundary condition; the absolute minimum limit of measure, beyond which the possibility for *persistence* itself is operationally cut off. In other words, this equation tautologically defines the minimum *boundary of time.*

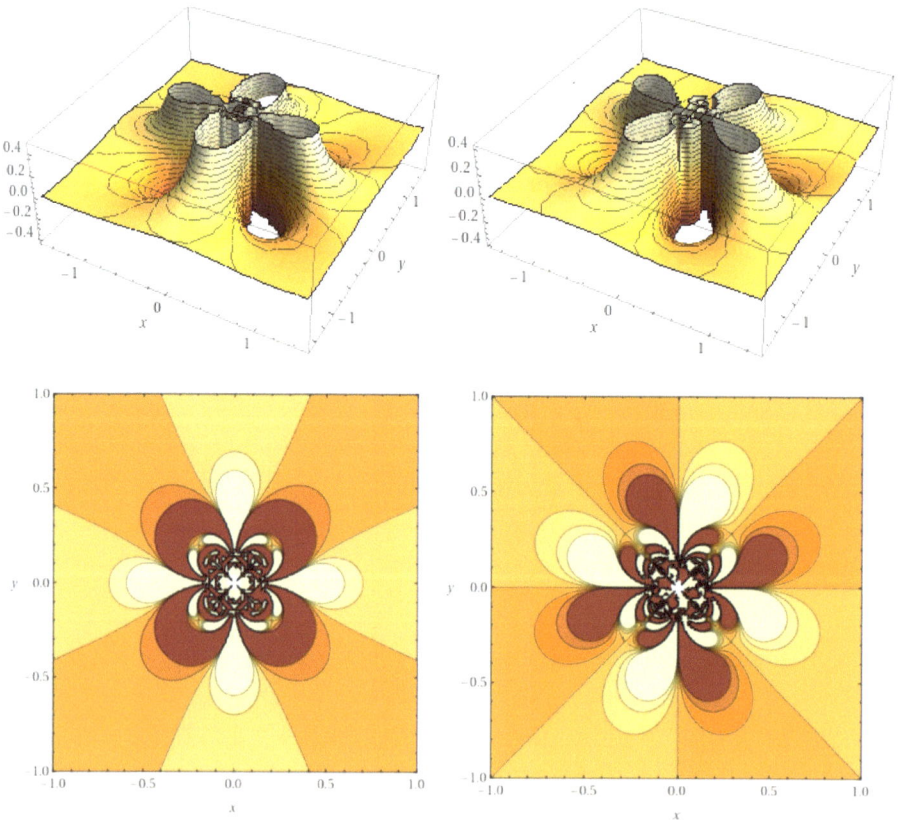

Graph 1: real (left) and imaginary (right) plots of the internal action of balance 0 under inverse-complex argument.

24

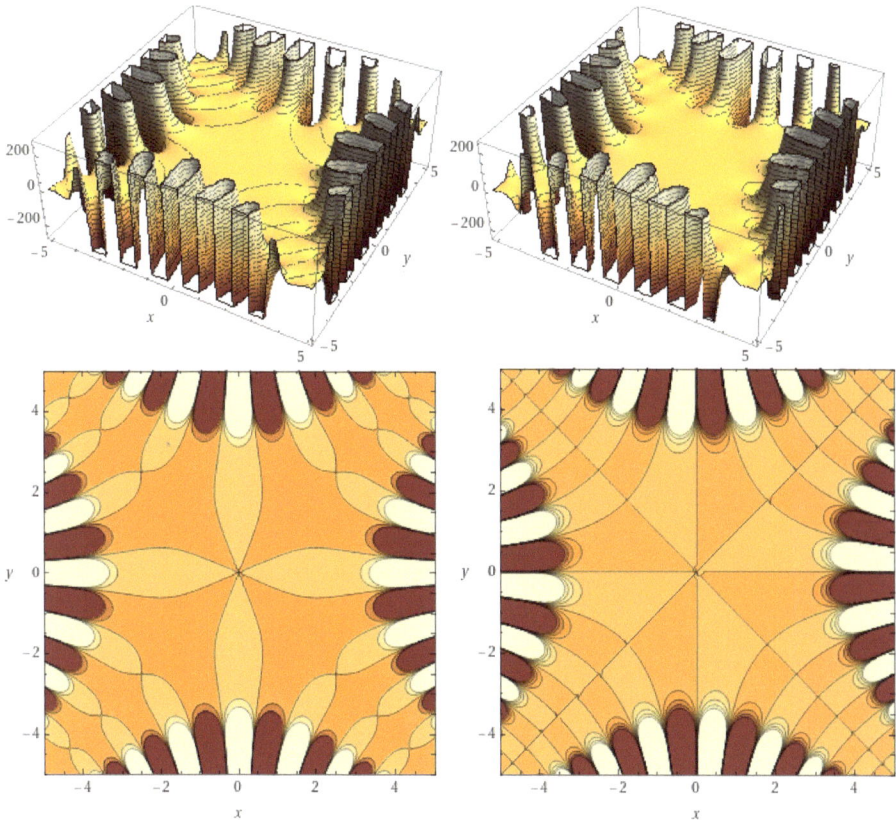

Graph 2: real (left) and imaginary (right) plots of the internal action of balance 0 under complex argument.

Equivalent expressions for the minimum limit of persistence (balance 0) include:

$$\pi \left(i\,sin\left(\left(\frac{\sqrt{i}}{2} \right)^2 \right) \right)^2 e^{\phi_0} = !\,\boldsymbol{n} \qquad\qquad \text{balance 0}$$

$$\frac{\pi}{2}\left(\frac{1}{2}\left(e^{1/2} + e^{-1/2} \right) - 1 \right) e^{\phi_0} = !\,\boldsymbol{n} \qquad\qquad \text{balance 0}$$

$$\frac{\pi}{2}\left(cosh\left(\frac{1}{2} \right) - 1 \right) e^{\phi_0} = !\,\boldsymbol{n} \qquad\qquad \text{balance 0}$$

$$cosh\left(\left(\frac{1}{2} \right)^2 \right)\left(sinh\left(\left(\frac{1}{2} \right)^2 \right) \right)^2 \left| \Gamma\left(\frac{1}{2} + \frac{i}{4\pi} \right) \right|^2 e^{\phi_0} = !\,\boldsymbol{n}$$

and

$$\cosh\left(\left(\frac{2\pi}{4\pi}\right)^2\right)\left(\sinh\left(\left(\frac{2\pi}{-4\pi}\right)^2\right)\right)^2\left|\Gamma\left(\frac{2\pi+i}{4\pi}\right)\right|^2 e^{\phi_0} = !n$$

Where $\pi =$ Archimedes' constant, $e =$ Euler's number, $n = 5$ the number of unique rotations partitioning the minimal arena, $!n =$ the derangement function, $sin(x) =$ the sine function, $sinh(x) =$ the hyperbolic sine function, $cosh(x) =$ the hyperbolic cosine function, $\Gamma(s) =$ the gamma function, $|x| =$ the absolute value function, and $i = \sqrt{-1}$ the imaginary unit.

Chapter 3: the next limit

The break in scale-symmetry between the first and second scales of any planar circular construction is $b = 7$. This follows from the fact that under planar circular construction (when we are constructing things from circles) the next circle available is composed of 7 circles (6 outer circles surrounding 1 inner circle) and the negative space that divides them.

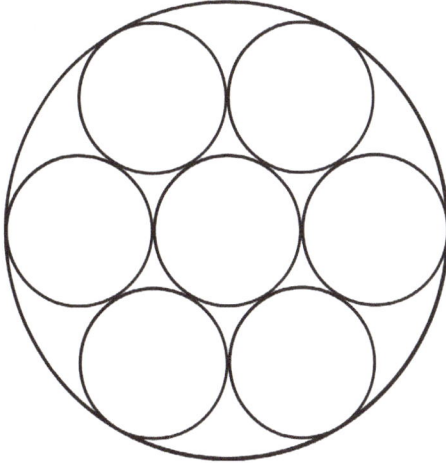

$$\frac{\text{\# of circles included on 2}^{\text{nd}} \text{ scale}}{\text{\# of circles included on 1}^{\text{st}} \text{ scale}} = b = 7$$

With this in mind, let's zoom out to the next partition limit of the minimal arena (balance 1), whose inner action is the operationally rotated inverse action of balance 0, adjusted for the break in scale symmetry between them. This balance maintains a new external rotation (ϕ_1) that rotates (cycles through) $d_1 = bn = 35$ of the 44 fundamental derangements.

$$\left(\sinh\left(\sinh\left(\frac{1}{b}\right)\right)\right)^{-1} e^{\phi_1} = d_1 \qquad\qquad \text{balance 1}$$

Where $\phi_1 = 1.61625918175645\ldots + 2\pi k\, i$ is the magnitude of the external rotation of balance 1, $\sinh(x) =$ the hyperbolic sine function, $b =$ the break in scale symmetry between the 2 internal balances, and $d_1 = 35$ the number of derangements participating in balance 1.

27

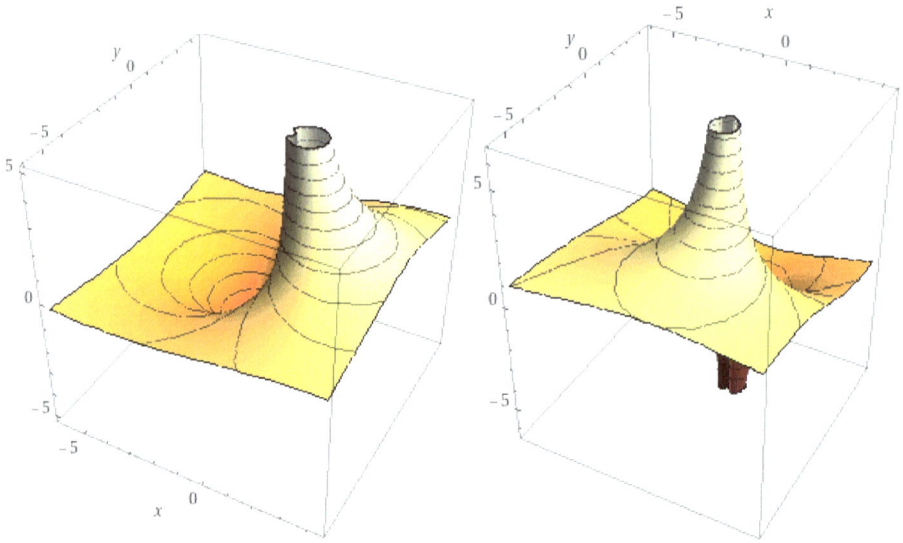

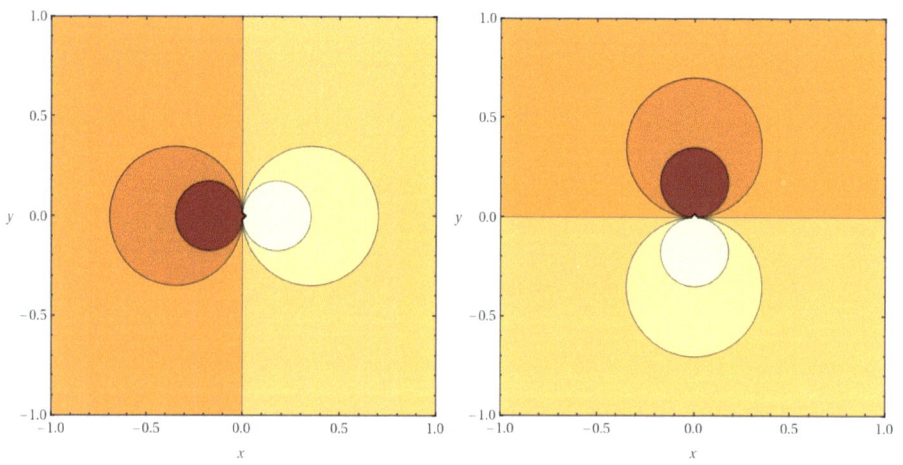

Graph 3: real (left) and imaginary (right) plots of the internal action of balance 1 under complex argument.

Balance 1 can be equivalently written:

$$\left(\frac{1}{2} \left(e^{\left(\frac{1}{2} \left(e^{1/b} - e^{-1/b} \right) \right)} - e^{\left(\frac{1}{2} \left(e^{-1/b} - e^{1/b} \right) \right)} \right) \right)^{-1} e^{\Phi_1} = d_1$$

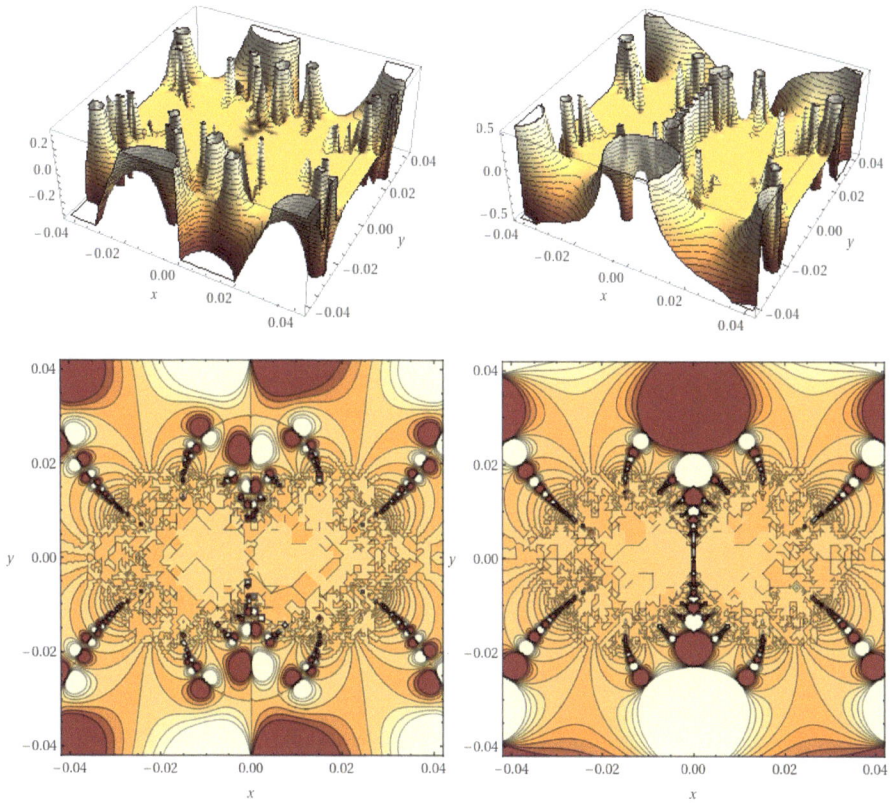

Graph 4: real (left) and imaginary (right) plots of the internal action of balance 1 under inverse-complex argument, zoomed in.

The **n** partition rotations of the minimal arena squarely connect (**n**²) *and* orthogonally hyperbolically factor (Γ(**n**)) under ideal hyperbolic logarithmic arrangement.

$$b\,cosh(\,\log b\,) = n^2$$

$$b\,sinh(\,\log b\,) = \Gamma(n)$$

Where $cosh(x) = $ the hyperbolic cosine function, $sinh(x) = $ the hyperbolic sine function, $\log(x) = $ the hyperbolic ("natural") logarithm function, $\Gamma(s) = $ the gamma function which encodes hyperbolically balanced partitions, $n = 5$ the number of unique rotations partitioning the minimal arena and, $b = 7$ the break in scale symmetry between the 2 internal balances.

Chapter 4: the external balances

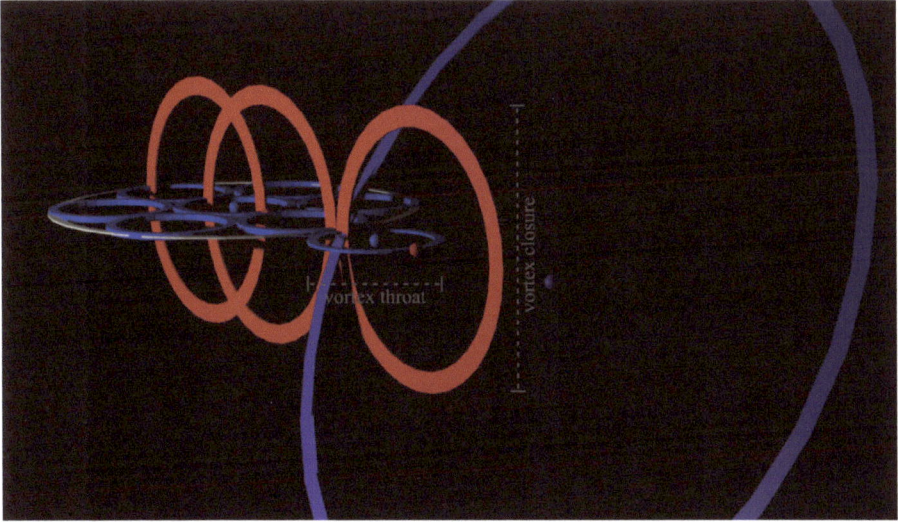

The external (purple) domain is maintained under closed hyperbolic connection. The blue and red balances within that domain define the vortex's throat (balance 2) and its closure (balance 3).

     The external domain of the minimal arena connects under ideal hyperbolic vortex arrangement (see Chapter 6). Within that connection balance 2 forms the vortex's throat, defining where the n unique partitions doubly-periodically divide (W_{We}), while factoring along the break in scale symmetry of the geometry $\sqrt{b}$, and wrapping axisymmetrically around a single pole $(3)^{1/3}$.

$$\frac{n}{W_{We} \sqrt{b}\ (3)^{1/3}}\, e^{\phi_2} = d_2 \qquad\qquad \text{balance 2}$$

Where $\phi_2 = 1.87554596713962\ldots + 2\pi k\, i$ is the magnitude of the external rotation of balance 2, $n = 5$ the number of unique rotations partitioning the minimal arena, $b = 7$ the break in scale symmetry between the 2 internal balances of that arena, $d_2 = 18$ the number of derangements participating in balance 2, $e =$ Euler's number, and $W_{We} =$ is the Weierstrass constant—the unitary balance of the Weierstrass sigma function.

The Weierstrass constant characterizes the n-dimensional hypersphere balanced over unitary lemniscate division.

$$W_{We} = \frac{1}{L}\left(\frac{e^\pi}{4}\right)^{1/8} = (2^{2n}\, e^\pi)^{1/8}\, \frac{\Gamma\left(\frac{1}{2}\right)}{\left(\Gamma\left(\left(\frac{1}{2}\right)^2\right)\right)^2}$$

$$i^{-i} = 2\, L^4\, W_{We}{}^4$$

$$L = \frac{1}{2}s \qquad\qquad s = \frac{1}{\sqrt{2\pi}}\left(\Gamma\left(\left(\frac{1}{2}\right)^2\right)\right)^2$$

Where $\pi =$ Archimedes' constant, $e =$ Euler's number, $L =$ the lemniscate constant, $i = \sqrt{-1}$ the imaginary unit, s is the arc length of the unitary lemniscate (with $a = 1$), $\Gamma(s) =$ the gamma function, $\Gamma\left(\frac{1}{2}\right) = \sqrt{\pi}$, and $e^\pi =$ the volume sum of all even-dimensional hyperspheres, which follows from the fact that the equations for the volume and surface areas of n-dimensional hyperspheres of radius r are:

$$V_n(r) = \frac{\pi^{n/2}}{\Gamma\left(\frac{n}{2}+1\right)}\, r^n \qquad\qquad S_{n-1}(r) = \frac{2\pi^{n/2}}{\Gamma\left(\frac{n}{2}\right)}\, r^{n-1}$$

$$\lim_{n\to\infty} \frac{\pi^0}{\Gamma(1)} + \frac{\pi^1}{\Gamma(2)} + \frac{\pi^2}{\Gamma(3)} + \cdots + \frac{\pi^n}{\Gamma(n+1)} = e^\pi$$

$$\lim_{n\to\infty} \frac{\pi^0}{0!} + \frac{\pi^1}{1!} + \frac{\pi^2}{2!} + \frac{\pi^3}{3!} + \cdots + \frac{\pi^n}{n!} = e^\pi$$

$$\Gamma(n) = (n-1)!$$

Where $\Gamma(n) =$ the gamma function, and $n!$ is the factorial function.

This partitioning has a few orthogonally balanced components. That is, in addition to being balanced by the unitary Weierstrass sigma function, the throat of the hyperbolic vortex is also orthogonally maintained by the Weierstrass elliptic functions which have equianharmonic, lemniscatic, and pseudo-lemniscatic half-periods.

equianharmonic half-periods

$$\omega_1 = \frac{\Gamma^3\left(\frac{1}{3}\right)}{4\pi}(-1)^{\frac{1}{3}} \qquad \omega_2 = \frac{\Gamma^3\left(\frac{1}{3}\right)}{4\pi}$$

lemniscatic half-periods

$$\frac{L}{2\sqrt{2}} \qquad\qquad \frac{L}{2\sqrt{2}}i$$

pseudo-lemniscatic half-periods

$$\frac{L}{4}(i+1) \qquad\qquad \frac{L}{4}(i-1)$$

The lemniscate plays a central (quadropoly orthogonal) role in this external balance because it defines "the inverse curve of the hyperbola with respect to its center." [4]

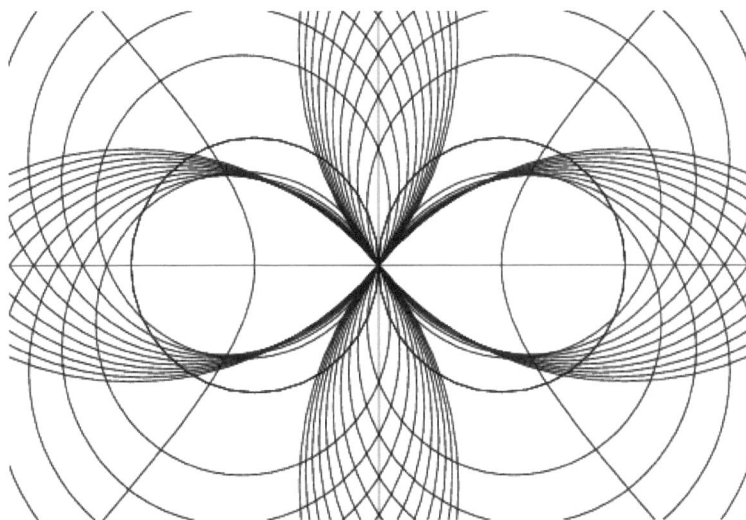

The lemniscate can also be generated as the envelope of circles centered on a rectangular hyperbola and passing through its center.

---

[4] Wells, D. *The Penguin Dictionary of Curious and Interesting Geometry.* London: Penguin, pp. 139-140, 1991. Figure and caption from mathworld.wolfram.com/Lemniscate.html

Instead of writing balance 2 in terms of the Weierstrass constant, or the lemniscate constant, let's write it in terms of the gamma function, to highlight how it is maintained under ideal square/split hyperbolic balance.

$$\frac{n}{\sqrt{b\pi}\ (3)^{1/3}}\left(2^{2n}\ e^{\pi}\right)^{-1/8}\left(\Gamma\left(\left(\frac{1}{2}\right)^2\right)\right)^2 e^{\phi_2} = d_2 \qquad \text{balance 2}$$

Where $\phi_2 = 1.87554596713962\ldots + 2\pi k\ i$ is the magnitude of the external rotation of balance 2, π = Archimedes' constant, e = Euler's number, $\Gamma(s)$ = the gamma function, which encodes hyperbolic partitions, $n = 5$ the number of unique rotations partitioning the minimal arena, and $d_2 = 18$ the number of derangements participating in balance 2.

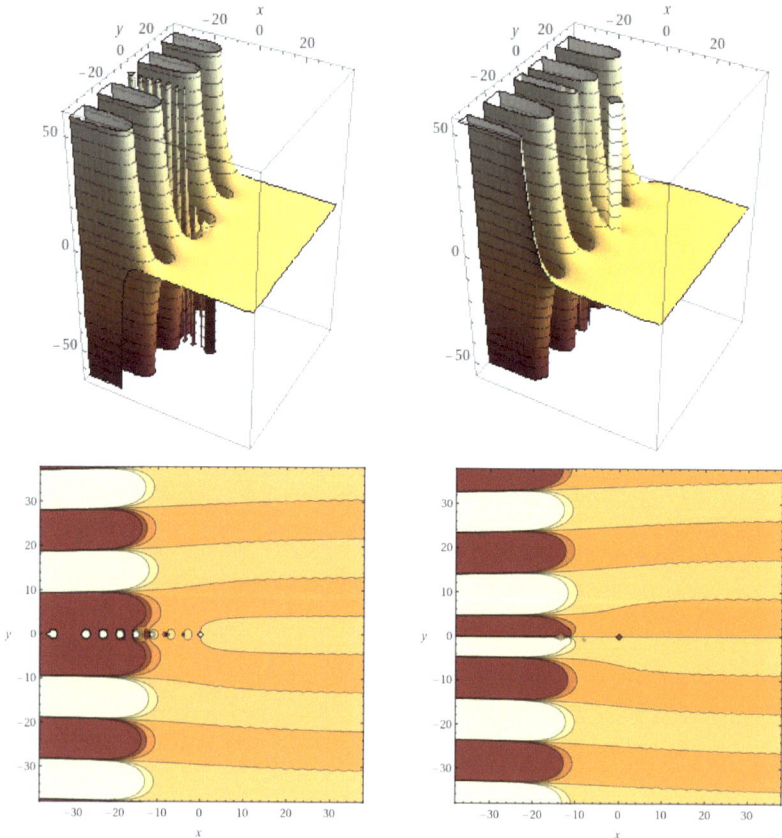

Graph 5: real (left) and imaginary (right) plots of the internal action of balance 2 under dual complex argument.

34

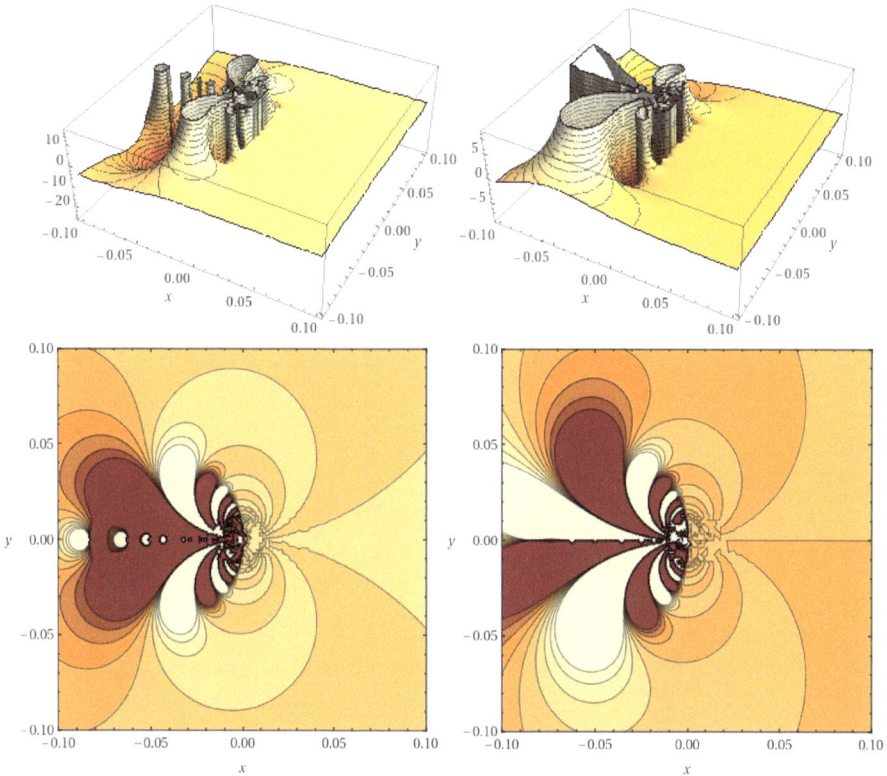

Graph 6: real (left) and imaginary (right) plots of the internal action of balance 2 under dual inverse-complex argument. "the partition butterfly"

Balance 3 defines the opposite end of the hyperbolic vortex, where the 5 rotations of the minimal arena circularly close ($2\pi\,n$) under square *connective circular balance*, forming split figure eights.

$$2\pi\,n\left(cos\left(\frac{b}{n}\right)\right)^2 e^{\phi_3} = d_3 \qquad \text{balance 3}$$

Where $\phi_3 = 2.17642683817579\ldots + 2\pi k\,i$ is the magnitude of the external rotation maintaining balance 3, π = Archimedes' constant, e = Euler's number, $n = 5$ the number of unique rotations partitioning the minimal arena, $b = 7$ the break in scale symmetry between the 2 internal balances of that arena, $cos\left(\frac{b}{n}\right)$ = the connective circular balance, and $d_3 = 8$ the number of derangements participating in balance 3.

35

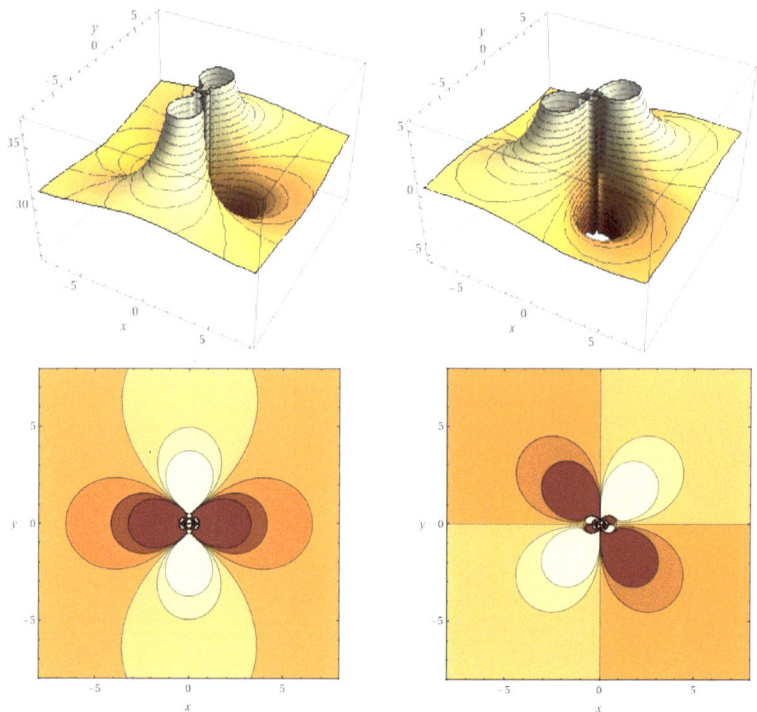

Graph 7: real (left) and imaginary (right) plots of the internal action of balance 3 under inverse-complex argument. "split figure eights"

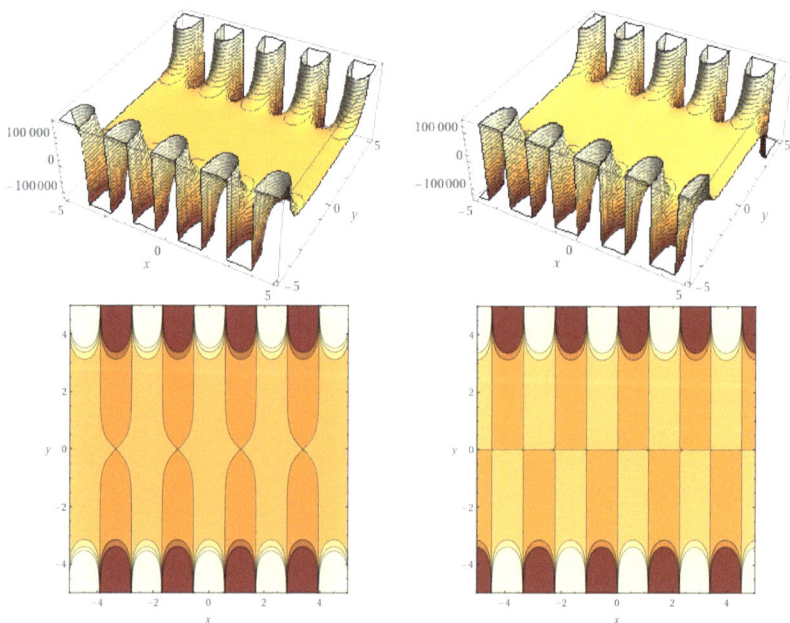

Graph 8: real (left) and imaginary (right) plots of the internal action of balance 3 under complex argument.

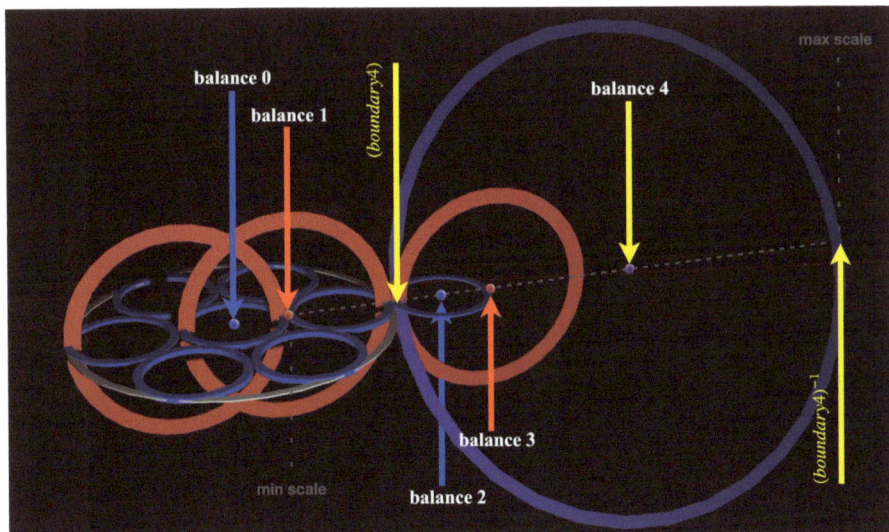

The internal and external boundaries of balance 4 (*boundary* 4 &
$(boundary\ 4)^{-1}$) *are under inverse action.*

Balance 4 defines the projective balance point of the minimal
arena—the minimal self-divisible unitary reference. In other words,
balance 4 = 1.

Boundary 4 (defined as the union of balance 2 and balance 3) is
the product of joined inverse factors $\left(\frac{1}{x} + x\right) = e^{\log x} + e^{-\log x} =$
$2\,cosh(\ \log x\ )$, the square *hyperbolic rotational split* of the minimal
arena, and its square connective circular balance.

$$2\,cosh(\log b)\left(cosh\left(\frac{n}{2}\right)\right)^2\left(cos\left(\frac{b}{n}\right)\right)^2 e^{\phi_4} = d_4 \qquad \text{boundary 4}$$

Where $\phi_4 = 1.41678698590795\ ... + 2\pi k\ i$ is the magnitude of the 4th
external rotation, e = Euler's number, $n = 5$ the number of unique
rotations partitioning the minimal arena, $b = 7$ the break in scale
symmetry between the 2 internal boundaries of that arena (and the
external logarithmic connection), $cosh(x)$ = the hyperbolic cosine
function, $cos(x)$ = the cosine function, $log(x)$ = the hyperbolic
logarithm function, $cosh\left(\frac{n}{2}\right)$ = the hyperbolic rotational split of the
minimal arena, $cos\left(\frac{b}{n}\right)$ = its connective circular balance, and $d_4 = 32$
the number of derangements maintaining boundary 4.

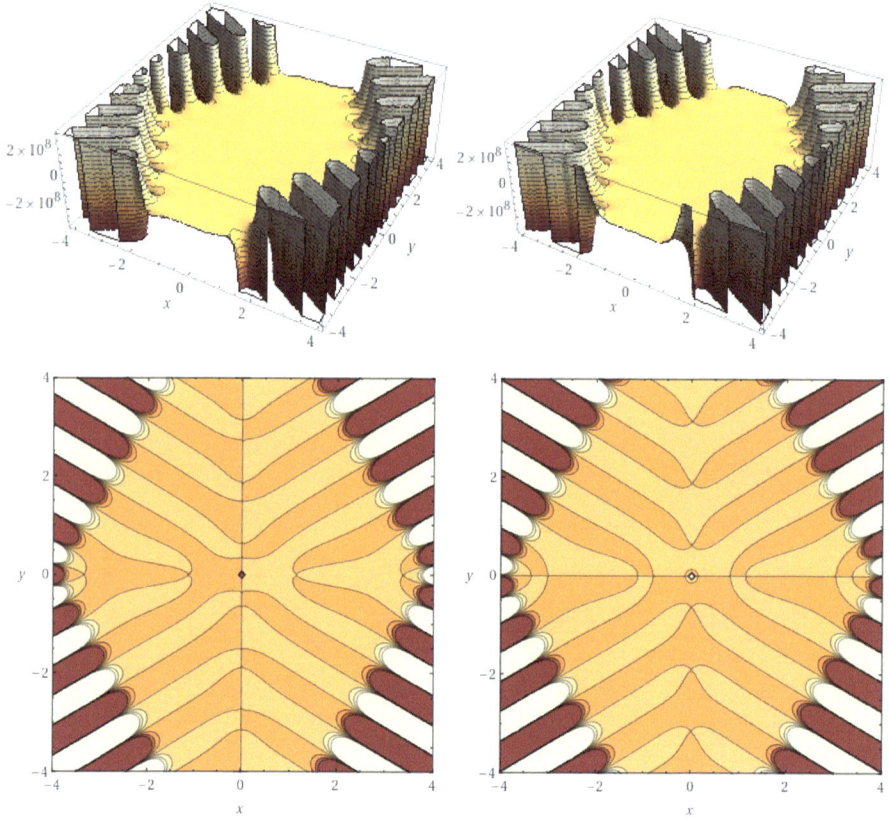

Graph 9: real (left) and imaginary (right) plots of the inner action of boundary 4 under triple complex argument.

(boundary 4)$^{-1}$ defines the outer edge of action in this minimally bounded universe, the other boundary which no action participating in the balance of persistence crosses.

$$(2\,cosh(\log b))^{-1} \left(cosh\left(\tfrac{n}{2}\right)\right)^{-2} \left(cos\left(\tfrac{b}{n}\right)\right)^{-2} e^{-\phi_4} = d_4^{-1} \quad \text{(boundary 4)}^{-1}$$

Where $\phi_4 = 1.41678698590795\ldots + 2\pi k\,i$ is the magnitude of the 4th boundary's external rotation, e = Euler's number, $n = 5$ the number of unique rotations partitioning the minimal arena, $b = 7$ the break in scale symmetry between the 2 interior balances of that arena, $cosh(x)$ = the hyperbolic cosine function, $cos(x)$ = the cosine function, $\log(x)$ = the hyperbolic logarithm function, and $d_4 = 32$ the derangements maintaining (boundary 4).

38

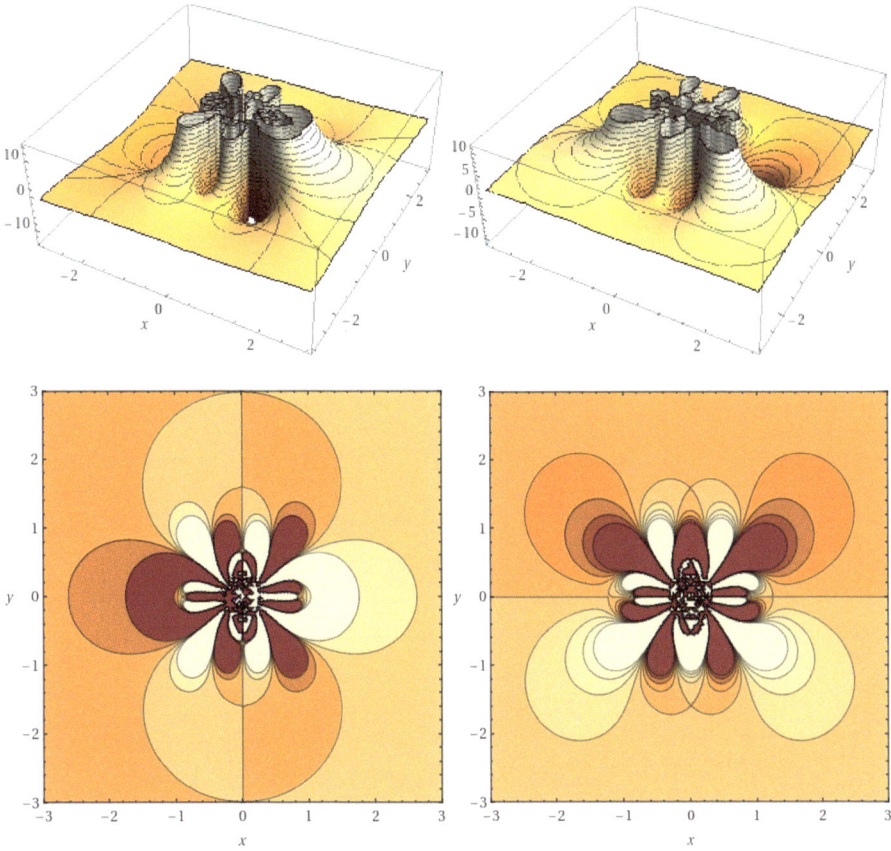

Graph 10: real (left) and imaginary (right) plots of the inner action of boundary 4 under triple inverse-complex argument.

Now that we have decomposed the minimal arena into its 5 unique balanced derangements, let's examine their boundary conditions.

decomposition map of the minimal arena

$$(2\cosh(\log b))^{-1}\left(\cosh\left(\frac{n}{2}\right)\right)^{-2}\left(\cos\left(\frac{b}{n}\right)\right)^{-2} e^{-\phi_4} = d_4^{-1} \quad (4^{th})^{-1}$$

$$(\text{external reference}) = 1$$

$$2\pi\, n\left(\cos\left(\frac{b}{n}\right)\right)^2 e^{\phi_3} = d_3 \qquad 3^{rd}$$

$$\frac{n}{\sqrt{b\pi}\,(3)^{1/3}}\,(2^{2n}\,e^{\pi})^{-1/8}\left(\Gamma\left(\left(\frac{1}{2}\right)^2\right)\right)^2 e^{\phi_2} = d_2 \qquad 2^{nd}$$

$$2\cosh(\log b)\left(\cosh\left(\frac{n}{2}\right)\right)^2\left(\cos\left(\frac{b}{n}\right)\right)^2 e^{\phi_4} = d_4 \qquad 4^{th}$$

$$\left(\sinh\left(\sinh\left(\frac{1}{b}\right)\right)\right)^{-1} e^{\phi_1} = d_1 \qquad 1^{st}$$

$$\pi\left(\sinh\left(\left(\frac{1}{2}\right)^2\right)\right)^2 e^{\phi_0} = d_0 \qquad 0^{th}$$

Where ϕ_k = the external rotation of the kth balance, d_k = the number of derangements participating in the kth balance, $n = 5$ the number of unique rotations maintained by the partition balance of the minimal arena, $!n = 44$ the number of derangements available to 5 rotations, $b = 7$ the break in scale symmetry between the internal 2 balances of that arena, π = Archimedes' constant, e = Euler's number, $\sinh(x)$ = the hyperbolic sine function, $\cosh(x)$ = the hyperbolic cosine function, $\cos(x)$ = the cosine function, $\log(x)$ = the hyperbolic/natural logarithm function, and $\Gamma(s)$ = the gamma function, which encodes hyperbolically balanced partitions.

$\phi_4 = 1.4167869859079\ldots + (2\pi k)i \quad d_4 = 2(\sqrt{d_0 - d_1} + 1)^2 = 32$

$\phi_3 = 2.1764268381757\ldots + (2\pi k)i \quad d_3 = 2(\sqrt{d_0 - d_1} - 1)^2 = 8$

$\phi_2 = 1.8755459671396\ldots + (2\pi k)i \quad d_2 = 2(\sqrt{d_0 - d_1} \pm 0)^2 = 18$

$\phi_1 = 1.6162591817564\ldots + (2\pi k)i \quad d_1 = bn = 35$

$\phi_0 = 5.3912583683231\ldots + (2\pi k)i \quad d_0 = \,!n = 44$

Chapter 5: the minimal arena's boundaries

With perfect precision, the partition boundaries of the minimal arena define the limits of measure for the 5 physical bases—time, space, charge, mass and temperature, known as the Planck constants. Each external rotation (ϕ_k) balanced as a double cover derangement defines a Planck boundary $Planck\ boundary_k = \phi_k(2n)^{\pm d_k}$.

Planck temperature	$T_p = \phi_4(2n)^{+d_4}$
Planck mass	$m_p = \phi_3(2n)^{-d_3}$
Planck charge	$q_p = \phi_2(2n)^{-d_2}$
Planck length	$l_p = \phi_1(2n)^{-d_1}$
Planck time	$t_p = \phi_0(2n)^{-d_0}$

$T_p = 1.41678698590709 \ldots \times 10^{32}\ K$	predicted
$T_p = 1.416784(16) \times 10^{32}\ K$	measured
$m_p = 2.1764268381757 \ldots \times 10^{-8}\ kg$	predicted
$m_p = 2.176434(24) \times 10^{-8}\ kg$	measured
$q_p = 1.8755459671396 \ldots \times 10^{-18}\ C$	predicted
$q_p = 1.8755459 \times 10^{-18}\ C$	previously defined
$l_p = 1.6162591817564 \ldots \times 10^{-35}\ m$	predicted
$l_p = 1.616255(18) \times 10^{-35}\ m$	measured
$t_p = 5.3912583683231 \ldots \times 10^{-44}\ s$	predicted
$t_p = 5.391247(60) \times (10)^{-44}\ s$	measured

Where the digits in the parentheses define the measurement error of the preceding two digits ($5.391247(60)$ means 5.391247 ± 0.000060), t_p, l_p, q_p, m_p, T_p = the Planck time, length, charge, mass, and temperature, $\phi_0, \phi_1, \phi_2, \phi_3, \phi_4$ = the 5 unique partition rotations dividing the minimal arena, d_0, d_1, d_2, d_3, d_4 = the number of derangements maintaining each of those rotations, and $n = 5$.

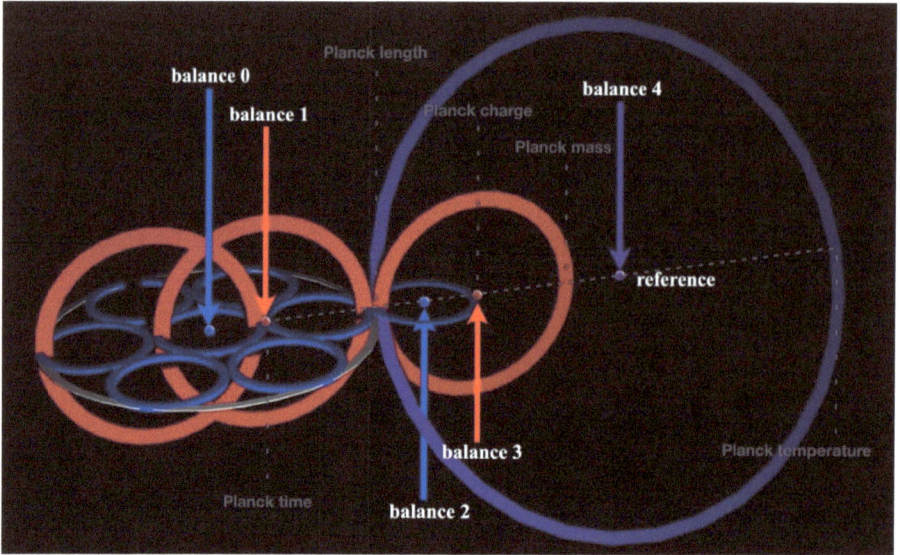

The Planck constants define the 5 partition limits (boundaries) of the minimal arena, defined internally as the minimum possible volume complement (the hyperbolic figure eight knot) and externally as the n-hypersphere of maximal volume.

Geometry's simplest possible projective arena defines the *coherent system of units*, and the bounds of that minimal system define the Planck constants (the boundaries of reality's 5 fundamental dimensions—time, space, charge, mass, and temperature).

Let's examine how the external domain of this arena connects.

Chapter 6: the hyperbolic vortex

Let the external domain of the minimal arena be maintained under self-closed hyperbolic connection.

$$\frac{1}{ж} + ж + ж^3 = \text{hyperbolic connection}$$

$$\frac{1}{ж} + ж + \frac{ж^3}{2\pi} = \text{self} - \text{closed hyperbolic connection}$$

Where ж is a variable.

And let the action responsible for this self-closed hyperbolic connection trivially involve 2 orthogonal twists, both of which are internal to the Planck mass boundary (the partition limit).

$$\left(i^i \right)^{-\frac{\pi}{2}} - m_p$$

Where $i = \sqrt{-1}$, $i^i = e^{-\frac{\pi}{2}}$ is a single twist (a rotation of $\frac{\pi}{2}$ radians), $\left(i^i \right)^{-\frac{\pi}{2}}$ is an orthogonal combination of 2 internal twists, and m_p = the Planck mass boundary.

Setting these two conditions equal to each other yields the hyperbolic vortex equation.

---

the hyperbolic vortex equation

$$\frac{1}{ж} + ж + \frac{ж^3}{2\pi} = \left(i^i \right)^{-\frac{\pi}{2}} - m_p$$

Where π = Archimedes' constant, $i = \sqrt{-1}$, $\left(i^i \right)^{-\frac{\pi}{2}}$ = an orthogonal combination of 2 internal twists of $\left(\frac{\pi}{2} \right)$ (quarter turns), and m_p = the Planck mass.

The hyperbolic vortex equation has 4 solutions composed of 6 parts (4 real and 2 imaginary).

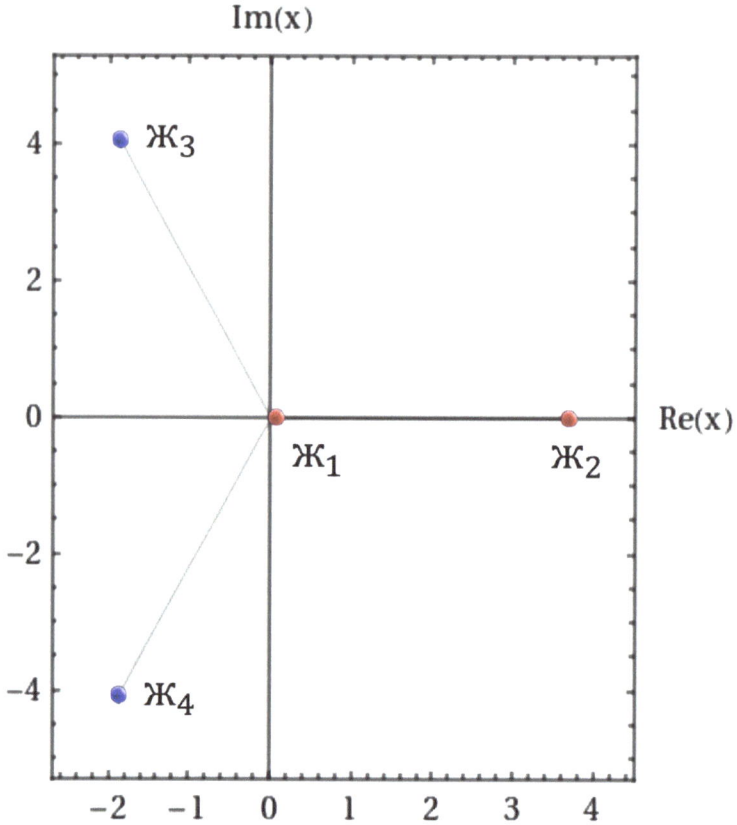

The 4 hyperbolic vortex partition constants.

$$Ж_1 = 0.0854245431533310\ldots$$
$$Ж_2 = 3.66756753485499\ldots$$
$$Ж_3 = -1.87649603900416\ldots + 4.06615262615971\ldots\, i$$
$$Ж_4 = -1.87649603900416\ldots - 4.06615262615971\ldots\, i$$

These partition constants are ideally cyclic and combinatorial. For example, their product is equal to 2π. They sum to zero, as do their real and imaginary parts. Their quadrance (the difference of their squares and the sum of their squares) is spherical ($\pm 4\pi$). And so on.

$$\text{ж}_1\text{ж}_2\text{ж}_3\text{ж}_4 = 2\pi \qquad\qquad \text{product}$$

$$\text{ж}_1 + \text{ж}_2 + \text{ж}_3 + \text{ж}_4 = 0 \qquad\qquad \text{sum}$$

$$\text{ж}_1 + \text{ж}_2 + Re(\text{ж}_3) + Re(\text{ж}_4) = 0 \qquad\qquad \text{real sum}$$

$$Im(\text{ж}_3) + Im(\text{ж}_4) = 0 \qquad\qquad \text{imaginary sum}$$

$$\text{ж}_1{}^2 + \text{ж}_2{}^2 + \text{ж}_3{}^2 + \text{ж}_4{}^2 = -4\pi \qquad\qquad \text{quadrance}$$

$$Re(\text{ж}_3) = Re(\text{ж}_4) = -\tfrac{1}{2}(\text{ж}_1 + \text{ж}_2) \qquad\qquad -\text{ joined twins}$$

$$Im(\text{ж}_3) = -Im(\text{ж}_4) = \tfrac{1}{2}(Im(\text{ж}_3) - Im(\text{ж}_4)) \qquad\qquad +\text{ split twins}$$

$$\text{ж}_3 + \text{ж}_4 = 2Re(\text{ж}_3) = 2Re(\text{ж}_4) \qquad\qquad \text{partial sum}$$

$$\text{ж}_3 - \text{ж}_4 = 2Im(\text{ж}_3) = 2Im(\text{ж}_4) \qquad\qquad \text{partial difference}$$

$$Re(\text{ж}_3)^2 + Re(\text{ж}_4)^2 = \tfrac{1}{2}(\text{ж}_1 + \text{ж}_2)^2 \qquad\qquad \text{real square sum}$$

$$Re(\text{ж}_3)^2 + Im(\text{ж}_3)^2 = Re(\text{ж}_4)^2 + Im(\text{ж}_4)^2 = \text{ж}_3\text{ж}_4 = \text{ж}_r{}^2 \quad \text{pair}$$

$$Re(\text{ж}_3)^2 - Re(\text{ж}_4)^2 = Im(\text{ж}_3)^2 - Im(\text{ж}_4)^2 = \text{ж}_3\text{ж}_4 = \text{ж}_r{}^2 \quad \text{pair}$$

$$Im(\text{ж}_3)^4 = Im(\text{ж}_4)^4 = Im(\text{ж}_3)^2\, Im(\text{ж}_4)^2 \qquad\qquad \text{imaginary}^4$$

$$Re(\text{ж}_3)^4 = Re(\text{ж}_4)^4 = \left(\left(-\tfrac{2\pi}{4\pi} \right)(\text{ж}_1 + \text{ж}_2) \right)^4 \qquad\qquad \text{real}^4$$

$$\left(\frac{\text{ж}_1{}^2 + \text{ж}_2{}^2 + \text{ж}_3{}^2 + \text{ж}_4{}^2}{\text{ж}_1 \times \text{ж}_2 \times \text{ж}_3 \times \text{ж}_4} \right) = -2 \qquad\qquad \text{self intersection number}$$

ж_3 and ж_4 can also be expressed in terms of radius and angle.

$\text{ж}_r = 4.47826244916751 \ldots$ hyperbolic vortex radius constant
$\text{ж}_\theta = 2.00316562310924 \ldots$ hyperbolic vortex radian constant

$$\text{ж}_3 = \text{ж}_r\, e^{\text{ж}_\theta i} \qquad\qquad\qquad \text{ж}_3\text{ж}_4 = \text{ж}_r{}^2$$

$$\text{ж}_4 = \text{ж}_r\, e^{-\text{ж}_\theta i} \qquad\qquad\qquad \text{ж}_1\text{ж}_2\text{ж}_r{}^2 = 2\pi$$

$$\text{Ж}_r = \sqrt{\text{Ж}_3 \text{Ж}_4} \qquad\qquad \log\left(\frac{\text{Ж}_3}{\text{Ж}_r}\right) = \text{Ж}_\theta i$$

$$\text{Ж}_1 + \text{Ж}_2 + \frac{\text{Ж}_r{}^2}{\text{Ж}_3} + \frac{\text{Ж}_r{}^2}{\text{Ж}_4} = 0 \qquad\qquad \log\left(\frac{\text{Ж}_4}{\text{Ж}_r}\right) = -\text{Ж}_\theta i$$

$$\sum_{k=1}^{4} \text{Ж}_k = 0 \qquad\qquad \prod_{k=1}^{4} \text{Ж}_k = 2\pi$$

$$\sum_{k=1}^{4} \text{Ж}_k{}^2 = -4\pi \qquad\qquad \prod_{k=1}^{4} \text{Ж}_k{}^2 = 4\pi^2$$

$$\frac{\text{Ж}_1 \text{Ж}_2 \text{Ж}_3 \text{Ж}_4}{4} = \frac{\pi}{2}$$

$$\frac{\pi}{2}\left(\sum_{k=1}^{4} \text{Ж}_k{}^2\right)^2 = \left(\prod_{k=1}^{4} \text{Ж}_k\right)^3$$

Combinatorics of the hyperbolic vortex's partition constants.

These partitions solutions define how the external facing boundaries of the minimal arena divide under closed hyperbolic connection. In other words, they define how the Planck *charge* and Planck *mass* boundaries naturally partition. Let's explore those partitions.

Chapter 7: charge partitions

Externally *charge* has a magnitude equal to the charge boundary (the Planck charge q_p) multiplied by the 1st hyperbolic vortex partition constant (Ж_1). This defines the charge of the electron (see Chapter 10).

$$e = \text{Ж}_1 q_p$$

Internally this *charge* boundary partitions precisely as the volume of the hyperbolic figure eight knot internally partitions (in the imaginary direction), into a negative one-thirds and a positive two-thirds division of a dilogarithmic flip $Li_2(-1)$.

$$V_{fe} = i\left[-Li_2\left((-1)^{\frac{1}{3}}\right) + Li_2\left(-(-1)^{\frac{2}{3}}\right)\right]$$

Where Li_2 is the hyperbolic dilogarithm (the polylogarithm of order 2).

This arranges *charge* into the following partitions.

charge partitions

(Z) (H) (W)	(± 1) (0) (± 1)
(P) (N) (e)	$(+1)$ (0) (-1)
(v_τ) (v_μ) (v_e)	(0) (0) (0)

external
- -
internal

(d) (e) (u)	$\left(-\frac{1}{3}\right)(-1)\left(+\frac{2}{3}\right)$
(s) (μ) (c)	$\left(-\frac{1}{3}\right)(-1)\left(+\frac{2}{3}\right)$
(b) (τ) (t)	$\left(-\frac{1}{3}\right)(-1)\left(+\frac{2}{3}\right)$

Where $e = \text{Ж}_1 q_p$ is set as the scale of charge unity $= 1$, and Z, H, W, P, N, e, v_τ, v_μ, v_e, d, u, s, μ, c, b, τ and t are the charge assignments of the Z boson, Higgs boson, W boson, proton, neutron, electron, tau neutrino, muon neutrino, electron neutrino, down quark, up quark, strange quark, muon, charm quark, beauty (bottom) quark, tau quark, and truth (top) quark.

In summary, the smallest external boundary of the minimal arena (the Planck charge boundary q_p defining the throat of the hyperbolic vortex) partitions into the charge assignments of *all* the fundamental particles of matter.

These charge partitions $(e, 0, -e, -\frac{1}{3}e, +\frac{2}{3}e)$ maintain the following combinatorial attributes.

The sum of *external* charges is 0.

The product of *external* charges is 0.

The sum of *external* unit charge magnitudes is 6.

The sum of *internal* unit charges is -2.

The sum of *internal* unit charge magnitudes is 6.

The product of *internal* unit charge magnitudes is equal to the derangements of the mass boundary squared divided by the derangements of the charge boundary cubed $\frac{d_3{}^2}{d_2{}^3}$.

And the sum of all internal and external charges is -2.

$$\sum q_{external} = 0 \qquad \prod q_{external} = 0$$

$$\sum |q_{external}| = 6 \qquad \sum q_{internal} = -2$$

$$\sum |q_{internal}| = 6 \qquad \prod q_{internal} = \frac{d_3{}^2}{d_2{}^3}$$

$$\sum_{k=1}^{18} q_k = -2$$

Where $e = \text{ж}_1 q_p$ is set as the scale of charge unity $= 1$, $|x| =$ the absolute value function, $d_2 =$ the 18 derangements of the charge boundary, and $d_3 =$ the 8 derangements of the mass boundary.

Chapter 8: mass partitions

The Planck mass boundary (m_p) defines the partition limit of the minimal arena. This limit has 3 orthogonal expressions defining: (1) how it balances, (2) how it splits, and (3) how it joins.

The *balance* of the mass boundary traces out two figure eight knot volumes ($2V_{fe}$) in 4 orthogonal ways (m_p^4) while inverse hyperbolically terminating the derangements ($!n$) of the minimal arena. This geometric arrangement of the mass boundary defines the electron mass (m_e).

---

balance map

$$m_e = 2V_{fe}\, m_p^4 \left(1 + \left(sinh\left(sinh\left(\frac{!n}{b}\right)\right)^{-1}\right)^{-1} \boxplus\right)$$

$$m_e = 9.10938370161994 \ldots \times 10^{-31}\, kg \qquad \text{predicted}$$
$$m_e = 9.1093837015(28) \times 10^{-31}\, kg \qquad \text{measured}$$

Where V_{fe} = the complement volume of the hyperbolic figure eight knot, m_p = the Planck mass boundary, $sinh(x)$ = the hyperbolic sine function, $n = 5$, $!n = 44$ the number of derangements available to 5 rotations, $b = 7$ the break in scale symmetry between the minimal arena's internal 2 balances, and $\boxplus = \left(\frac{l_p\, m_p}{q_p^2}\right)$ is the terminal boundary of that arena.

Every externally balanced partition action within this ideal geometry terminates on this universal binomial division boundary $\boxplus = \left(\frac{l_p\, m_p}{q_p^2}\right)$, the partition boundary of the *n*-hypersphere of *maximal* volume.

$$\sum_{k=0}^{4} \phi_k = \left(\frac{d_3}{d_2}\right)\frac{4\pi}{Re(\omega_1)^3}\left(1 + \left(\frac{d_2}{d_3}\right)\frac{V_h^*}{n}\boxplus\right)$$

Where π = Archimedes' constant, $Re(\omega_1)$ = the Real part of the omega_1 constant (the equianharmonic half-period from Chapter 4), V_h^* = the dimension at which an *n*-hypersphere has *maximal* volume, $n = 5$, $\sum \phi_k$ = the sum of unique rotations partitioning the minimal arena, d_2 = the 18 derangements of the charge boundary, d_3 = the 8 derangements of the mass boundary, and $\boxplus$ = the terminal boundary.

Any squarely balanced (periodic) space has a total of 17 possible tessellations, or tilings. (There are exactly 17 plane symmetry groups.) To find all 17 tessellations of the Planck *mass* boundary we extend its trivial balance (m_e) under ideal (splitting) ^{2} and orthogonal joining— guaranteeing that:

1. the Planck charge and Planck mass boundaries partition under hyperbolic vortex connection ($ж_1, ж_2, ...$),
2. that connection squarely hyperbolically bifurcates (splits) in 2 orthogonal but otherwise indistinguishable ways ($\alpha_F - 1$)^2,
3. and it dilogarithmically persists under a unitary hyperbolic-circular closed balance (being logarithmically (μ, γ) braided under the $- \left(\frac{1}{3} \right)$ and $+ \left(\frac{2}{3} \right)$ internal split action of the hyperbolic figure eight knot).

How this partition limit (the Planck mass boundary) *splits* defines the proton mass (m_+). And how it orthogonally *joins* defines the neutron mass (m_N).

---

splitting & joining maps

$$\left(\frac{m_e}{m_+} \right) \left(\frac{ж_2}{ж_1} \right)^2 = \left(\frac{2}{3} \right)^2 (\alpha_F - 1)^2 \left(1 - \left(\frac{1}{3} \right) e^{3\gamma} \boxplus \right)$$

$$\left(\frac{m_N - m_+}{m_e} \right) = \left(\frac{1}{3} \right) (\mu + 3 + \pi) \left(1 + \left(\frac{2}{3} \right) e^{3\gamma} \boxplus \right)$$

$m_+ = 1.67262192371195 ... \times 10^{-27} \, kg$	predicted
$m_+ = 1.67262192369(51) \times 10^{-27} \, kg$	measured
$m_N = 1.67492749802284 ... \times 10^{-27} \, kg$	predicted
$m_N = 1.67492749804(95) \times 10^{-27} \, kg$	measured

Where m_e = the electron mass, m_+ = the proton mass, m_N = the neutron mass, π = Archimedes' constant, e = Euler's number, α_F = the alpha Feigenbaum constant (the 2nd period-doubling bifurcation constant of the logistic map), μ = the nontrivial zero of the logarithmic integral γ = the Euler-Mascheroni constant (the limiting difference between the harmonic series and the natural logarithm function), $ж_1$ & $ж_2$ = the 1st and 2nd hyperbolic vortex partition constants, and $\boxplus$ = the terminal boundary of the minimal arena.

50

Mirroring the division structure of the minimal arena's internal complement volume (the hyperbolic figure eight knot), which arranges 6 circles around a central circle that successively undergoes 4 orthogonal internal bifurcations, the mass boundary's trivial joining map $\left(\frac{m_N - m_+}{m_e} \right)$ extends into a balance of externally phased and internally folded tessellations. And the orthogonally connected splitting map $\left(\frac{m_e}{m_+} \right) \left(\frac{\text{ж}_2}{\text{ж}_1} \right)^2$ of the mass boundary extends into a balance of twists, splits and flips.

$$= \left(\frac{m_H - m_Z}{m_W} \right) 6$$

$$= \left(\frac{m_N - m_+}{m_e} \right)$$

$$= \left(\frac{m_d - m_u}{m_e} \right) \left(\frac{1}{2} \right)$$

$$= \left(\frac{m_c - 2m_s}{m_\mu} \right) \left(\frac{1}{2} \right)^2$$

$$= \left(\frac{m_t - m_\tau}{m_b} \right) \left(\left(\frac{1}{2} \right)^2 \right)^2$$

$$= \left(\frac{m_{\nu_\tau} - m_{\nu_\mu}}{m_{\nu_e}} \right) \left(\left(\left(\frac{1}{2} \right)^2 \right)^2 \right)^2$$

Note: Since the masses of the neutrinos have not been measured, unveiling the neutrinos requires completing the patterns below, using the fact that the mass boundary is symmetric and closed.

full joining map

$$\left(\frac{m_H - m_Z}{m_W}\right) = \left(\frac{1}{3}\right)^2 (\mu + 3 + \pi)(2^{-1})\left(1 + 6\left(\frac{Ж_r}{W(1)}\right)^4 \boxplus\right)$$

$$\left(\frac{m_N - m_+}{m_e}\right) = \left(\frac{1}{3}\right)(\mu + 3 + \pi)(2^0)\left(1 + \left(\frac{2}{3}\right)e^{3\gamma} \boxplus\right)$$

$$\left(\frac{m_d - m_u}{m_e}\right) = \left(\frac{1}{3}\right)(\mu + 3 + \pi)(2^{+1})\left(1 - \left(\frac{b}{n}\right)\sqrt{S_S}\,\frac{Ж_r^4}{4}\boxplus\right)$$

external phases

- -

internal folds

$$\left(\frac{m_c - 2m_s}{m_\mu}\right) = \left(\frac{1}{3}\right)(\mu + 3 + \pi)(2^2)\left(1 - \left(\frac{3}{V_h^*}\right)^3 \frac{Ж_r^4}{2}\boxplus\right)$$

$$\left(\frac{m_t - m_\tau}{m_b}\right) = \left(\frac{1}{3}\right)(\mu + 3 + \pi)\left(2^{2^2}\right)\left(1 + \left(\frac{b}{6}\right)D_{DHA}^2\,Ж_r^2 \boxplus\right)$$

$$\left(\frac{m_{\nu_\tau} - m_{\nu_\mu}}{m_{\nu_e}}\right) = \left(\frac{1}{3}\right)(\mu + 3 + \pi)\left(2^{2^{2^2}}\right)\left(1 - ???\boxplus\right)$$

Where m_e, m_+, m_N, m_μ, m_τ, m_Z, m_W, m_H, m_t, m_c, m_b, m_s, m_d, m_u, m_{ν_τ}, m_{ν_μ} and m_{ν_e} are respectively the masses of the: electron, proton, neutron, muon, tau, Z boson, W boson, Higgs boson, truth (top) quark, charm quark, beauty (bottom) quark, strange quark, down quark, up quark, tau neutrino, muon neutrino, and the electron neutrino, $n = 5$, $b = 7$, π = Archimedes' constant, μ = the Ramanujan-Soldner constant (the nontrivial zero of the logarithmic integral), e = Euler's number, γ = the Euler-Mascheroni constant (which defines the limiting difference between the natural logarithm and the harmonic series), $Ж_r$ = the hyperbolic vortex radius constant, V_h^* = the dimension at which an n-hypersphere has *maximal* volume, $W(1)$ = the omega constant (the unique real number that satisfies the equation $xe^x = 1$), D_{DHA} = the offset at which 2 unit disks overlap by half each's area, S_S = the Silverman constant, and $\boxplus$ = the terminal boundary of the minimal arena.

full splitting map

$$\frac{m_b + m_c + m_t}{\left(\sqrt{m_b} + \sqrt{m_c} + \sqrt{m_t}\right)^2} = \left(\frac{2}{3}\right)^3 (\alpha_F - 1)^2 \left(1 - \sqrt{W_{We}}\left(2\,Im(\text{ж}_3)\right)^4 \boxplus\right)$$

$$\left(\frac{m_e}{m_+}\right)\left(\frac{\text{ж}_2}{\text{ж}_1}\right)^2 = \left(\frac{2}{3}\right)^2 (\alpha_F - 1)^2 \left(1 - \left(\frac{1}{3}\right)e^{3\gamma} \boxplus\right)$$

$$\frac{m_e + m_\mu + m_\tau}{\left(\sqrt{m_e} + \sqrt{m_\mu} + \sqrt{m_\tau}\right)^2} = \left(\frac{2}{3}\right)^1 (\alpha_F - 1)^0 \left(1 + 2L^2\left(Re(\text{ж}_3)\right)^3 \boxplus\right)$$

- -

$$\frac{m_H + m_Z + m_W}{\left(\sqrt{m_H} + \sqrt{m_Z} + \sqrt{m_W}\right)^2} = \left(\frac{2}{3}\right)^{-\frac{1}{2}} (\alpha_F - 1)^{-e^{2\gamma}} \left(1 + \left(\frac{3}{n}\right)\sqrt{P_{up}}^{\,n}\,\text{ж}_r^{\,2} \boxplus\right)$$

$$\frac{m_u + m_s + m_d}{\left(\sqrt{m_u} + \sqrt{m_s} + \sqrt{m_d}\right)^2} = \left(\frac{2}{3}\right)^0 \gamma\,(\alpha_F - 1)^0 \left(1 - 6\left(2V_{fe}\right)\text{ж}_r^{\,3} \boxplus\right)$$

$$\frac{m_{\nu_e} + m_{\nu_\mu} + m_{\nu_\tau}}{\left(\sqrt{m_{\nu_e}} + \sqrt{m_{\nu_\mu}} + \sqrt{m_{\nu_\tau}}\right)^2} = \left(\frac{2}{3}\right)^{-\frac{1}{2}} (\alpha_F - 1)^{+e^{\pi i}} \left(1 + ??? \boxplus\right)$$

Where m_e, m_+, m_N, m_μ, m_τ, m_Z, m_W, m_H, m_t, m_c, m_b, m_s, m_d, m_u, m_{ν_τ}, m_{ν_μ} and m_{ν_e} are respectively the masses of the: electron, proton, neutron, muon, tau, Z boson, W boson, Higgs boson, truth (top) quark, charm quark, beauty (bottom) quark, strange quark, down quark, up quark, tau neutrino, muon neutrino, and the electron neutrino, V_{fe} = the complement volume of the hyperbolic figure eight knot, α_F = the alpha Feigenbaum constant, e = Euler's number, γ = the Euler-Mascheroni constant (which denotes the limiting difference between the natural logarithm and the harmonic series), ж_1, ж_2, ж_3, and ж_4 = the 1st, 2nd, 3rd, and 4th hyperbolic vortex partition constants, ж_r = the hyperbolic vortex radius constant, π = Archimedes' constant, P_{up} = the universal parabolic constant, W_{We} = the Weierstrass constant, L = the lemniscate constant, d_2 = the 18 derangements of the charge boundary, $n = 5$, and $\boxplus$ = the terminal boundary of the minimal arena.

These mass assignments couple as hyperbolic partition limits.

electron mass (m_e)

$$\frac{d_0}{d_3}\left(\frac{m_e}{m_b}\right) = 4\pi \, \text{Ж}_1{}^4 \qquad \frac{d_0}{n}\left(\frac{m_e}{m_\tau}\right) = 2V_{fe}\, \text{Ж}_1{}^3$$

$$\frac{2}{3}\left(\frac{m_e}{m_W}\right) = \frac{\text{Ж}_1{}^4}{4\pi} \qquad \left(\frac{m_e}{m_\mu}\right) = \frac{\text{Ж}_1{}^2}{\sinh(C_{CFP})}\left(1+\left(\frac{d_0\,d_1}{4\pi}\right)\boxplus\right)$$

Where π = Archimedes' constant, Ж_1 = the 1st hyperbolic vortex partition constant, V_{fe} = the volume complement of the figure eight hyperbolic knot, $n = 5$, $\sinh(x)$ = the hyperbolic sine function, C_{CFP} = the fixed point of the hyperbolic cotangent, d_0 = the 44 derangements of the time boundary, d_1 = the 35 derangements of the space boundary, d_3 = the 8 derangements of the mass boundary, m_b = the beauty (bottom) quark mass, m_τ = the tau mass, m_W = the W boson mass, and m_μ = the muon mass.

proton mass (m_+)

$$\frac{1}{\sqrt{3n}}\left(\frac{m_+}{m_s}\right) = P_{up} \qquad d_1 d_2\left(\frac{m_+}{m_z}\right) = 2\,\varphi\,\text{Ж}_\theta$$

Where P_{up} = the universal parabolic constant, φ = the golden ratio, $n = 5$, Ж_θ = the hyperbolic vortex radian constant, d_1 = the 35 derangements of the space boundary, d_2 = the 18 derangements of the charge boundary, m_s = the strange quark mass, and m_z = the Z boson mass.

neutron mass (m_N)

$$\frac{d_1}{d_2}\left(\frac{m_N}{m_H}\right) = 2\,\text{Ж}_1{}^2 \qquad \frac{1}{d_3}\left(\frac{m_N}{m_c}\right) = 4\pi\,\text{Ж}_1{}^2$$

Where π = Archimedes' constant, Ж_1 = the 1st hyperbolic vortex partition constant, d_1 = the 35 derangements of the space boundary, d_2 = the 18 derangements of the charge boundary, d_3 = the 8 derangements of the mass boundary, m_H = the Higgs mass, and m_c = the charm quark mass.

54

muon mass (m_μ)

$$\left(\frac{m_e}{m_\mu}\right) = \frac{\text{Ж}_1{}^2}{sinh(C_{CFP})}\left(1+\left(\frac{d_0\,d_1}{4\pi}\right)\boxplus\right)$$

Where π = Archimedes' constant, $\text{Ж}_1{}^2 = \alpha$ the fine-structure constant, $sinh(x)$ = the hyperbolic sine function, C_{CFP} = the fixed point of the hyperbolic cotangent, d_0 = the 44 derangements of the time boundary, and d_1 = the 35 derangements of the space boundary, and m_e = the electron mass.

$m_\mu = 1.88353163790140 \ldots \times 10^{-28}\ kg$ predicted
$m_\mu = 1.883531627(42) \times 10^{-28}\ kg$ measured

Z boson mass (m_Z)

$$d_1 d_2\left(\frac{m_+}{m_Z}\right) = 2\,\varphi\,\text{Ж}_\theta$$

Where φ = the golden ratio, and Ж_θ = the hyperbolic vortex radian constant, d_1 = the 35 derangements of the space boundary, d_2 = the 18 derangements of the charge boundary, and m_+ = the proton mass.

$m_Z = 1.62556312860926 \ldots \times 10^{-25}\ kg$ predicted
$m_Z = 1.625566(38) \times 10^{-25}\ kg$ measured

W boson mass (m_W)

$$\frac{2}{3}\left(\frac{m_e}{m_W}\right) = \frac{\text{Ж}_1{}^4}{4\pi} \qquad \frac{3\,d_0}{2\,d_3}\left(\frac{m_W}{m_b}\right) = (4\pi)^2$$

Where π = Archimedes' constant, Ж_1 = the 1st hyperbolic vortex partition constant, d_0 = the 44 derangements of the time boundary, d_3 = the 8 derangements of the mass boundary, m_e = the electron mass, and m_b = the bottom quark mass.

$m_W = 1.43310146854387 \ldots \times 10^{-25}\ kg$ predicted
$m_W = 1.43288(21) \times 10^{-25}\ kg$ measured

tau mass (m_τ)

$$\frac{d_0}{n}\left(\frac{m_e}{m_\tau}\right) = 2V_{fe}\,ж_1{}^3$$

Where V_{fe} = the volume complement of the hyperbolic figure eight knot, $ж_1$ = the 1st hyperbolic vortex partition constant, $n = 5$, d_0 = the 44 derangements of the time boundary, and m_e = the electron mass.

$m_\tau = 3.16754377503113 \ldots \times 10^{-27}\ kg$ predicted
$m_\tau = 3.16754(21) \times 10^{-27}\ kg$ measured

Higgs mass (m_H)

$$\frac{d_2}{d_1 d_3}\left(\frac{m_H}{m_c}\right) = 2\pi \qquad\qquad \frac{d_1}{d_2}\left(\frac{m_N}{m_H}\right) = 2\,ж_1{}^2$$

Where π = Archimedes' constant, $ж_1$ = the 1st hyperbolic vortex partition constant, d_1 = the 35 derangements of the space boundary, d_2 = the 18 derangements of the charge boundary, d_3 = the 8 derangements of the mass boundary, m_c = the charm quark mass, and m_N = the neutron mass.

$m_H = 2.23149658442420 \ldots \times 10^{-25}\ kg$ predicted
$m_H = 2.2315(28) \times 10^{-25}\ kg$ measured

charm quark mass (m_c)

$$\frac{d_2}{d_1 d_3}\left(\frac{m_H}{m_c}\right) = 2\pi \qquad\qquad \frac{1}{d_3}\left(\frac{m_N}{m_c}\right) = 4\pi\,ж_1{}^2$$

Where π = Archimedes' constant, $ж_1$ = the 1st hyperbolic vortex partition constant, d_1 = the 35 derangements of the space boundary, d_2 = the 18 derangements of the charge boundary, d_3 = the 8 derangements of the mass boundary, m_H = the Higgs mass, and m_N = neutron mass.

$m_c = 2.28313100509581 \ldots \times 10^{-27}\ kg$ predicted
$m_c = 2.272(63) \times 10^{-27}\ kg$ measured

truth (top) quark mass (m_t)

$$d_2{}^2 \left(\frac{m_N}{m_t} \right) = \frac{(4\pi)^2}{\text{Ж}_r{}^3}$$

Where π = Archimedes' constant, Ж_r = the hyperbolic vortex radius constant, d_2 = the 18 derangements of the charge boundary, and m_N = neutron mass.

$m_t = 3.08638372655297 \ldots \times 10^{-25} \; kg$ predicted
$m_t = 3.084(07) \times 10^{-25} \; kg$ measured

up quark mass (m_u)

$$\left(\frac{m_\tau}{m_u} \right) = \left(2\pi \, \text{Ж}_r \right)^2$$

Where π = Archimedes' constant, Ж_r = the hyperbolic vortex radius constant, and m_τ = the tau mass.

$m_u = 4.00077202715006 \ldots \times 10^{-30} \; kg$ predicted
$m_u = 3.92(89) \times 10^{-30} \; kg$ measured

strange quark mass (m_s)

$$\frac{1}{\sqrt{3n}} \left(\frac{m_+}{m_s} \right) = P_{up}$$

Where P_{up} = the universal parabolic constant, $n = 5$, and m_+ = the proton mass.

$m_s = 1.88130136459184 \ldots \times 10^{-28} \; kg$ predicted
$m_s = 1.69(16) \times 10^{-28} \; kg$ measured

beauty (bottom) quark mass (m_b)

$$\frac{d_0}{d_3}\left(\frac{m_e}{m_b}\right) = 4\pi\,\text{ж}_1{}^4 \qquad \frac{3\,d_0}{2\,d_3}\left(\frac{m_W}{m_b}\right) = (4\pi)^2 \qquad \left(\frac{m_\mu}{m_b}\right) = 6\,\text{ж}_1{}^2\,\bar{s}_{lse}{}^4$$

Where π = Archimedes' constant, ж_r = the hyperbolic vortex radius constant, $\bar{s}_{lse}$ = the mean line-between-square edges length (in hypercube line picking) d_0 = the 44 derangements of the time boundary, d_3 = the 8 derangements of the mass boundary, m_e = the electron mass, m_W = the W boson mass, m_μ = the muon mass.

$m_b = 7.48705738029755 \ldots \times 10^{-27}\ kg$ predicted
$m_b = 7.45(07) \times 10^{-27}\ kg$ measured

down quark mass (m_d)

$$\left(\frac{m_c}{m_d}\right) = \sqrt{\frac{\bar{s}_{lse}}{2}}\,\text{ж}_r{}^4$$

$m_d = 8.61183215072197 \ldots \times 10^{-30}\ kg$ predicted
$m_d = 8.3(07) \times 10^{-30}\ kg$ measured

tau neutrino mass (m_{ν_τ})

$m_{\nu_\tau} = ???\ kg$???
$m_{\nu_\tau} = ???\ kg$ no current measurement

muon neutrino mass (m_{ν_μ})

$m_{\nu_\mu} = ???\ kg$???
$m_{\nu_\mu} = ???\ kg$ no current measurement

electron neutrino mass (m_{ν_e})

$m_{\nu_e} = ???\ kg$???
$m_{\nu_e} = ???\ kg$ no current measurement

partition parameters of the minimal arena

$n = 5$ — number of unique rotations

$b = 7$ — break in scale symmetry

$d_0 = \,!n = 44$ — (full derangement) derangements of balance 0

$d_1 = bn = 35$ — derangements of balance 1

$d_2 = 2(\sqrt{d_0 - d_1} \pm 0)^2 = 18$ — derangements of balance 2

$d_3 = 2(\sqrt{d_0 - d_1} - 1)^2 = 8$ — derangements of balance 3

$d_4 = 2(\sqrt{d_0 - d_1} + 1)^2 = 32$ — derangements of boundary 4

$\phi_0 = 5.39125836832313\ldots + (2\pi k)i \quad k \in \mathbb{Z}$ — 0th external rotation

$\phi_1 = 1.61625918175645\ldots + (2\pi k)i$ — 1st external rotation

$\phi_2 = 1.87554596713962\ldots + (2\pi k)i$ — 2nd external rotation

$\phi_3 = 2.17642683817579\ldots + (2\pi k)i$ — 3rd external rotation

$\phi_4 = 1.41678698590795\ldots + (2\pi k)i$ — 4th external rotation

$t_P = 5.39125836832313\ldots \times 10^{-44}\ s$ — Planck time

$l_P = 1.61625918175645\ldots \times 10^{-35}\ m$ — Planck length

$q_P = 1.87554596713962\ldots \times 10^{-18}\ C$ — Planck charge

$m_P = 2.17642683817579\ldots \times 10^{-8}\ kg$ — Planck mass

$T_P = 1.41678698590795\ldots \times 10^{32}\ K$ — Planck temperature

$G_{Gi} = 1.01494160640965\ldots$ — Gieseking's constant

$V_{fe} = 2.02988321281930\ldots$ — figure eight knot complement volume

$e = 2.71828182845904\ldots$ — Euler's number

$\pi = 3.14159265358979\ldots$ — Archimedes' constant

$P_{up} = 2.29558714939263\ldots$ — universal parabolic constant

$L = 2.622057554292119\ldots$ — lemniscate constant

$Ж_1 = 0.0854245431533304\ldots$ — 1st hyperbolic vortex partition constant

$Ж_2 = 3.66756753485499\ldots$ — 2nd hyperbolic vortex partition constant

$Ж_3 = -1.87649603900417\ldots + 4.06615262615972\ldots i$ — 3rd hvpc

$Ж_4 = -1.87649603900417\ldots - 4.06615262615972\ldots i$ — 4th hvpc

$Ж_r = 4.47826244916751\ldots$ — hyperbolic vortex radius constant

$Ж_\theta = 2.00316562310924\ldots$ — hyperbolic vortex radian constant

$\varphi = 1.61803398874989\ldots$ — the golden ratio

$\gamma = 0.577215664901532\ldots$ — Euler-Mascheroni constant

$\bar{s}_{lse} = 0.869009055274534\ldots$ — mean line-between-square-edges length

$\alpha_F = 2.50290787509589\ldots$ — alpha Feigenbaum constant

$\delta_F = 4.66920160910299\ldots$ — delta Feigenbaum constant

$W_{We} = 0.474949379987920\ldots$ — Weierstrass constant

$\mu = 1.45136923488338\ldots$ — nontrivial zero of the logarithmic integral

$\omega_1 = 0.764977018528596\ldots + 1.32497062714087\ldots i$ omega_1 constant

$\omega_2 = 1.529954037057192\ldots$ omega_2 constant

$W(1) = 0.567143290409783\ldots$ the omega constant

$C_{CFP} = 1.19967864025773\ldots$ Real fixed point of the hyperbolic cotangent

$L_{LL} = 0.662743419349181\ldots$ Laplace limit

$A_h{}^* = 7.25694640486057\ldots$ dimension of maximal n-hypersphere area

$V_h{}^* = 5.25694640486057\ldots$ dimension of maximal n-hypersphere volume

$S_S = 1.78657645936592\ldots$ Silverman constant

$D_{DHA} = 0.807945506599034\ldots$ the offset of 2 unit disks overlapping by half

17 mass partitions

$m_e = 9.10938370161994\ldots \times 10^{-31}\ kg$ electron mass

$m_+ = 1.67262192371195\ldots \times 10^{-27}\ kg$ proton mass

$m_N = 1.67492749802284\ldots \times 10^{-27}\ kg$ neutron mass

$m_\mu = 1.88353163790140\ldots \times 10^{-28}\ kg$ muon mass

$m_Z = 1.62556312860926\ldots \times 10^{-25}\ kg$ Z boson mass

$m_W = 1.43310146854387\ldots \times 10^{-25}\ kg$ W boson mass

$m_\tau = 3.16754377503113\ldots \times 10^{-27}\ kg$ tau mass

$m_H = 2.23149658442420\ldots \times 10^{-25}\ kg$ Higgs boson mass

$m_t = 3.08638372655297\ldots \times 10^{-25}\ kg$ truth (top) quark mass

$m_c = 2.28313100509581\ldots \times 10^{-27}\ kg$ charm quark mass

$m_b = 7.48705738029755\ldots \times 10^{-27}\ kg$ beauty (bottom) quark mass

$m_s = 1.88130136459184\ldots \times 10^{-28}\ kg$ strange quark mass

$m_u = 4.00077202715006\ldots \times 10^{-30}\ kg$ up quark mass

$m_d = 8.61183215072197\ldots \times 10^{-30}\ kg$ down quark mass

$m_{\nu_\tau} = ???\ldots \times 10^{-??}\ kg$ tau neutrino mass

$m_{\nu_\mu} = ???\ldots \times 10^{-??}\ kg$ muon neutrino mass

$m_{\nu_e} = ???\ldots \times 10^{-??}\ kg$ electron neutrino mass

Where the black digits represent previously known values and green digits represent extended predictions.

    In summary, the smallest external boundary of the minimal arena (the Planck charge boundary) partitions into the charge assignments of the fundamental particles of matter. And its external partition limit (the Planck mass boundary) partitions into the mass assignments of the fundamental particles of matter.

Chapter 9: the constants of Nature

So far, we have discovered that the minimal arena (internally defined as the minimum possible volume complement, the hyperbolic figure eight knot, externally counter-balanced as the n-hypersphere of maximal volume) decomposes into 5 unique bases (time, space, charge, mass and temperature) with sharply defined boundaries (the Planck constants); and that the external *charge* and *mass* boundaries of that arena partition into the exact charge and mass values that define the fundamental particles of matter. In this chapter, we notice that the minimal arena's 44 unique derangements define the constants of Nature.

To unveil the geometry of each constant of Nature, we think of them as binomial combinations of a primary partition action A_p on a primary intersection of boundaries B_p, and a terminal partition action A_t on the minimal arena's terminal boundary $= \boxplus$.

---

universal binomial factorization prescription

$$\left(\; constant \; of \; Nature \; \right) = A_p B_p \left(\; 1 \pm A_t \; \boxplus \; \right)$$

Since the terminal action of this balance doesn't get expressed until the 7th digit, this equation can be solved in pieces. That is, since $\boxplus = \left(\frac{l_p \, m_p}{q_p{}^2} \right) = 1 \times 10^{-7}$ we can safely ignore the terminal contribution of this balance at first, solve for its primary action and primary boundary $A_p B_p$, and then return to the full equation to solve for the terminal action A_t.

Step 1
Temporarily set $A_t = 0$, and construct the primary boundary B_p from the dimensions of the constant, selecting from the minimal arena's primary partition boundaries $\{ \, t_p, l_p, (q_p, e), (m_p, m_e, m_+, m_N), T_p \, \}$.

Step 2
If $A_p \neq 1$, then solve for it in terms of the primary partition parameters $(\, \pi, \text{Ж}_1, \text{Ж}_r \,)$.

Step 3

Plug the solution for $A_p B_p$ back into the universal binomial factorization prescription and solve for the terminal action A_t of that constant of Nature in terms of the minimal arena's derangements d_0, d_1, d_2, d_3, d_4 and its terminal partition parameters $\pi, \text{Ж}_2, \text{Ж}_3, \text{Ж}_4, \text{Ж}_r, \text{Ж}_\theta$.

Let's work through 2 examples.

To find the primary intersection of boundaries for any constant of Nature we look at its balance of dimensions. For example, since the speed of light (c) has dimensions *meters* divided by *seconds*, its primary boundary B_p will be equal to the natural boundary of length (the Planck length) divided by the natural boundary of time (the Planck time).

speed of light $\qquad\qquad c = 2.99792458 \times 10^8 \dfrac{m}{s}$

$$B_p = \left(\frac{l_p}{t_p} \right) = 2.99792566287110 \ldots \times 10^8 \frac{m}{s}$$

This number matches the measured value for the speed of light to the first ~7 digits, therefore, $A_p = 1$. To finish, we plug $A_p B_p$ back into the universal binomial factorization prescription and solve for the terminal action of the speed of light: $A_t = - \left(\frac{2\pi}{d_1} \right) \text{Ж}_r{}^2$.

$$c = \left(\frac{l_p}{t_p} \right) \left(1 - \left(\frac{2\pi}{d_1} \right) \text{Ж}_r{}^2 \; \boxplus \; \right)$$

Where $\pi =$ Archimedes' constant, $\text{Ж}_r =$ the hyperbolic vortex radius constant, and $d_1 =$ the 35 derangements of the space boundary.

Following the same procedure for the Hartree energy E_h, we begin by ignoring its terminal contribution and matching its dimensions to the minimal arena's partition boundaries. Since its dimensions are meters squared multiplied by kilograms divided by seconds squared $m^2 kg/s^2$ we set its primary boundary equal to the Planck length squared multiplied by *the electron mass* divided by the Planck time squared.

Hartree energy $$E_h = 4.3597447222071(85) \times 10^{-18} \frac{m^2 kg}{s^2}$$

$$B_p = \left(\frac{l_p{}^2 \, m_e}{t_p{}^2} \right)$$

Next, we solve for the primary action in terms of the external domain's partition parameters: $A_p = ж_1{}^4$. Then we plug $A_p B_p$ back into the universal binomial factorization prescription and solve for the terminal partition action of this balance: $A_t = -\left(\frac{b}{n} \right) \frac{e^\pi}{ж_r}$.

$$E_h = ж_1{}^4 \left(\frac{l_p{}^2 \, m_e}{t_p{}^2} \right) \left(1 - \left(\frac{b}{n} \right) \frac{e^\pi}{ж_r} \boxplus \right)$$

Where $n = 5$ the number of unique rotations partitioning the minimal arena, and $b = 7$ the break in scale symmetry between the minimal arena's internal 2 balances.

---

balance template

$$A_p B_p \left(1 \pm A_t \boxplus \right) \qquad \boxplus = \left(\frac{l_p \, m_p}{q_p{}^2} \right)$$

constant of Nature:
electric constant

$$\varepsilon_0 = \frac{1}{4\pi} \left(\frac{t_p{}^2 \, q_p{}^2}{l_p{}^3 \, m_p} \right) \left(1 - \left(\frac{\pi}{d_2} \right) ж_r \boxplus \right)$$

Every constant of Nature is a binomial union between a primary geometric action on a primary arrangement of boundaries (the construction of which is determined by the dimensions of the constant) and a terminal geometric action on $\boxplus$ = the terminal boundary of the minimal arena.

This universal binomial factorization prescription allows us to geometrically unveil every constant of Nature.

63

fine-structure constant

$$\alpha = {\text{Ж}_1}^2$$

$\alpha = 7.29735257295522\ldots \times 10^{-3}$ predicted
$\alpha = 7.2973525698(24) \times 10^{-3}$ measured

elementary charge

$$e = \text{Ж}_1 q_p$$

$e = 1.60217657405973\ldots \times 10^{-19}\ C$ predicted
$e = 1.602176565(35) \times 10^{-19}\ C$ measured

gravitational coupling

$$\alpha_G = \left(\frac{m_e}{m_p}\right)^2$$

$\alpha_G = 1.75182147492404\ldots \times 10^{-45}$ predicted
$\alpha_G = 1.7518(21) \times 10^{-45}$ measured

magnetic constant

$$\mu_0 = 4\pi \boxplus \left(1 + 4\pi \left(\frac{\pi}{2}\right)^{-1} \boxplus\right)$$

$\mu_0 = 1.25663706140197\ldots \times 10^{-6}\ m\ kg/C^2$ predicted
$\mu_0 = 1.256637061\ldots \times 10^{-6}\ m\ kg/C^2$ previously defined

Schwinger magnetic induction

$$S_{mi} = \frac{1}{\text{Ж}_1}\left(\frac{m_e{}^2}{t_p\, q_p\, m_p}\right)\left(1 - \left(3\, Im\left(i^{i^{i^{\cdot^{\cdot}}}}\right)\right)^2 2\pi \boxplus\right)$$

$S_{mi} = 4.41899541452692\ldots \times 10^9\ kg/s\ C$ predicted
$S_{mi} = 4.419 \times 10^9\ kg/s\ C$ measured

von Klitzing constant

$$R_K = \frac{2\pi}{Ж_1{}^2}\left(\frac{l_p{}^2\, m_p}{t_p\, q_p{}^2}\right)\left(1 + \frac{1}{2}Re\left(i^{i^{i^{\cdot^{\cdot}}}}\right)Ж_r{}^2\ \boxplus\right)$$

Where $Re\left(i^{i^{i^{\cdot^{\cdot}}}}\right)$ = the real part of the infinite power tower of the imaginary unit $i = \sqrt{-1}$, $Ж_1{}^2 = \alpha$ the fine structure constant, and $Ж_r =$ the hyperbolic vortex radius constant ($Ж_1 Ж_2 Ж_r{}^2 = 2\pi$).

$R_K = 2.5812807449400712 \ldots \times 10^4\ m^2 kg/s\ C^2$ predicted
$R_K = 2.58128074434(84) \times 10^4\ m^2 kg/s\ C^2$ measured

quantized Hall conductance

$$H_C = \frac{Ж_1{}^2}{2\pi}\left(\frac{t_p\, q_p{}^2}{l_p{}^2 m_p}\right)\left(1 - \frac{1}{2}Re\left(i^{i^{i^{\cdot^{\cdot}}}}\right)Ж_r{}^2\ \boxplus\right)$$

$H_C = 3.87404586641820 \ldots \times 10^{-5}\ s\ C^2/m^2 kg$ predicted
$H_C = 3.87404614(17) \times 10^{-5}\ s\ C^2/m^2 kg$ measured

magnetic flux constant

$$\Phi_0 = \frac{\pi}{Ж_1}\left(\frac{l_p{}^2\, m_p}{t_p\, q_p}\right)\left(1 + sec^2\left(\left(\frac{1}{2}\right)^2\right)Ж_r\ \boxplus\right)$$

$\Phi_0 = 2.06783384793306 \ldots \times 10^{-15}\ m^2 kg/s\ C$ predicted
$\Phi_0 = 2.067833848 \times 10^{-15}\ m^2 kg/s\ C$ previously defined

Josephson constant

$$K_J = \frac{Ж_1}{\pi}\left(\frac{t_p\, q_p}{l_p{}^2\, m_p}\right)\left(1 - sec^2\left(\left(\frac{1}{2}\right)^2\right)Ж_r\ \boxplus\right)$$

$K_J = 4.8359784854056 \ldots \times 10^{14}\ s\ C/m^2 kg$ predicted
$K_J = 4.835978484 \times 10^{14}\ s\ C/m^2 kg$ previously defined

quantum of circulation

$$q_c = \pi \left(\frac{l_p{}^2 m_p}{t_p m_e} \right) \left(1 + \sqrt{\zeta(2)} \; \text{ж}_\theta{}^2 \; \boxplus \right)$$

Where $\zeta(s)$ = the Reimann zeta function, ж_θ = the hyperbolic vortex radian constant, and $\boxplus$ = the terminal boundary of the minimal arena.

$q_c = 3.63694755140949 \ldots \times 10^{-4} \; m^2/s$ predicted
$q_c = 3.6369475516(11) \times 10^{-4} \; m^2/s$ measured

conductance quantum

$$G_0 = \frac{\text{ж}_1{}^2}{\pi} \left(\frac{t_p \, q_p{}^2}{l_p{}^2 m_p} \right) \left(1 - \sqrt{\zeta(3)} \; \text{ж}_\theta{}^2 \; \boxplus \right)$$

Where $\text{ж}_1{}^2 = \alpha$ the fine structure constant, and $\text{ж}_1 q_p = e$.

$G_0 = 7.74809172907834 \ldots \times 10^{-5} \; s \, C^2/m^2 kg$ predicted
$G_0 = 7.748091729 \times 10^{-5} \; s \, C^2/m^2 kg$ previously defined

2nd radiation constant

$$c_2 = 2\pi \left(l_p \, T_p \right) \left(1 - \sinh^2(C_{CFP}) \, \text{ж}_r{}^2 \; \boxplus \right)$$

Where C_{CFP} = the real fixed point of the hyperbolic cotangent, $\coth(C_{CFP}) = C_{CFP}$, $\sinh(C_{CFP}) = \frac{1}{L_{LL}}$, and L_{LL} = the Laplace limit.

$c_2 = 1.43877687654906 \ldots \times 10^{-2} \; m \, K$ predicted
$c_2 = 1.438776877 \times 10^{-2} \; m \, K$ previously defined

Planck's constant

$$\hbar = \left(\frac{l_p{}^2 m_p}{t_p} \right) \left(1 + \left(\frac{4\pi}{\sinh^2(2)} \right) \text{ж}_r \; \boxplus \right)$$

Where $\sinh(x)$ = the hyperbolic sine function.

$\hbar = 1.05457172603011 \ldots \times 10^{-34} \; m^2 kg/s$ predicted
$\hbar = 1.054571726(47) \times 10^{-34} \; m^2 kg/s$ measured

Coulomb's constant

$$\kappa = \left(\frac{l_p{}^3\, m_p}{t_p{}^2\, q_p{}^2} \right) \left(1 + \left(\frac{\pi}{d_2} \right) \text{ж}_r \ \boxplus \right)$$

$\kappa = 8.98755179228829 \dots \times 10^9 \ m^3 kg/s^2 C^2$ predicted
$\kappa = 8.9875517923(14) \times 10^9 \ m^3 kg/s^2 C^2$ measured

electric constant

$$\varepsilon_0 = \frac{1}{4\pi} \left(\frac{t_p{}^2\, q_p{}^2}{l_p{}^3\, m_p} \right) \left(1 - \left(\frac{\pi}{d_2} \right) \text{ж}_r \ \boxplus \right)$$

$\varepsilon_0 = 8.85418781277322 \dots \times 10^{-12} \ s^2 C^2/m^3 kg$ predicted
$\varepsilon_0 = 8.8541878128(13) \times 10^{-12} \ s^2 C^2/m^3 kg$ measured

the speed of light

$$c = \left(\frac{l_p}{t_p} \right) \left(1 - \left(\frac{2\pi}{d_1} \right) \text{ж}_r{}^2 \ \boxplus \right)$$

Where π = Archimedes' constant, d_1 = the 35 derangements of the space boundary, and ж_r = the hyperbolic vortex radius constant.

$c = 2.99792458354727 \dots \times 10^8 \ m/s$ predicted
$c = 2.99792458 \times 10^8 \ m/s$ previously defined

Faraday constant

$$F = N_A\, \text{ж}_1\, q_p \left(1 + \left(\frac{n}{d_0} \right) \frac{\text{ж}_r{}^2}{2\pi} \ \boxplus \right)$$

Where ж_r = the hyperbolic vortex radius constant ($\text{ж}_1 \text{ж}_2 \text{ж}_r{}^2 = 2\pi$), $n = 5$, d_0 = the 44 derangements of the time boundary, N_A = Avogadro's number, a *mol* combines the dimensions of $q_p\, m_p$, and $\text{ж}_1 q_p = e$ the electron charge.

$F = 9.64853321242908 \dots \times 10^4 \ C/mol$ predicted
$F = 9.648533212 \times 10^4 \ C/mol$ previously defined

neutron g-factor

$$g_N = -\frac{\Gamma(n)}{2\pi}\left(\frac{m_N}{m_+}\right)\left(1 + n\left(\frac{d_2}{4\pi}\right)\text{ж}_r{}^4 \; \boxplus\right)$$

Where $n = 5$, π = Archimedes' constant, ж_r = the hyperbolic vortex radius constant, d_2 = the 18 derangements of the charge boundary, m_N = the neutron mass, m_+ = the proton mass, and $\Gamma(s)$ = the gamma function.

$g_N = -3.82608560204984\dots$ predicted
$g_N = -3.82608545(90)$ measured

gravitational constant

$$G = \left(\frac{l_p{}^3}{t_p{}^2 m_p}\right)\left(1 - \left(\frac{\pi}{2}\right)\Gamma(n)\,\text{ж}_r{}^2 \; \boxplus\right)$$

Where $n = 5$, π = Archimedes' constant, ж_r = the hyperbolic vortex radius constant, and $\Gamma(s)$ = the gamma function.

$G = 6.67384038951738\dots \times 10^{-11}\ m^3/s^2kg$ predicted
$G = 6.67384(80) \times 10^{-11}\ m^3/s^2kg$ measured

Boltzmann constant

$$k_B = \left(\frac{l_p{}^2 m_p}{t_p{}^2 T_p}\right)\left(1 - \left(\frac{\pi}{2}\right)^2\text{ж}_r{}^2 \; \boxplus\right)$$

$k_B = 1.38064931695409\dots \times 10^{-23}\ m^2kg/s^2K$ predicted
$k_B = 1.380649 \times 10^{-23}\ m^2kg/s^2K$ previously defined

Compton angular frequency

$$\omega_C = \left(\frac{m_e}{t_p m_p}\right)\left(1 - \frac{1}{2}\Gamma(n)\; \boxplus\right)$$

$\omega_c = 7.7634409944556\dots \times 10^{20}\ 1/s$ predicted
$\omega_c = 7.763441 \times 10^{20}\ 1/s$ measured

spectral radiance constant

$$c_{1L} = 4\pi \left(\frac{l_p{}^4 m_p}{t_p{}^3} \right) \left(1 - \left(\frac{d_3}{d_1} \right) 4\pi \; \boxplus \right)$$

Where π = Archimedes' constant, d_1 = the 35 derangements of the space boundary, d_3 = the 8 derangements of the mass boundary.

$c_{1L} = 1.19104287769811 \ldots \times 10^{-16} \; m^4 kg/s^3$ predicted
$c_{1L} = 1.191042869(53) \times 10^{-16} \; m^4 kg/s^3$ measured

1st radiation constant

$$c_1 = (2\pi)^2 \left(\frac{l_p{}^4 m_p}{t_p{}^3} \right) \left(1 - \left(\frac{d_3}{d_1} \right) V_{fe} \; \text{Ж}_r \; \boxplus \right)$$

Where V_{fe} = the volume complement of the hyperbolic figure eight knot.

$c_1 = 3.74177185197781 \ldots \times 10^{-16} \; m^4 kg/s^3$ predicted
$c_1 = 3.741771852 \times 10^{-16} \; m^4 kg/s^3$ previously defined

characteristic impedance

$$Z_0 = 4\pi \left(\frac{l_p{}^2 m_p}{t_p \, q_p{}^2} \right) \left(1 + \left(\frac{d_1}{2} \right) \frac{\sqrt{G_{Gi}}}{\text{Ж}_\theta{}^2} \; \boxplus \right)$$

Where π = Archimedes' constant, G_{Gi} = Gieseking's constant for the minimum 3-manifold, Ж_θ = the hyperbolic vortex radian constant, d_1 = the 35 derangements of the space boundary, and $\boxplus$ = the terminal boundary of the minimal arena.

$Z_0 = 3.76730313666885 \ldots \times 10^2 \; m^2 kg/s \, C^2$ predicted
$Z_0 = 3.76730313668(57) \times 10^2 \; m^2 kg/s \, C^2$ measured

Bohr electron radius

$$a_0 = \frac{1}{Ж_1{}^2}\left(\frac{l_p\, m_p}{m_e}\right)\left(1 + \left(\frac{d_1}{d_3}\right)Ж_\theta \boxplus\right)$$

Where $Ж_1$ = the 1st hyperbolic vortex partition constant, and $Ж_\theta$ = the hyperbolic vortex radian constant.

$a_0 = 5.29177210889649 \ldots \times 10^{-11} \; m$ predicted
$a_0 = 5.2917721092(17) \times 10^{-11} \; m$ measured

classical electron radius

$$r_e = Ж_1{}^2\left(\frac{l_p\, m_p}{m_e}\right)\left(1 + \left(\frac{d_1}{d_3}\right)2 \boxplus\right)$$

$r_e = 2.81794032505462 \ldots \times 10^{-15} \; m$ predicted
$r_e = 2.8179403227(19) \times 10^{-15} \; m$ measured

Hartree energy

$$E_h = Ж_1{}^4\left(\frac{l_p{}^2\, m_e}{t_p{}^2}\right)\left(1 - \left(\frac{b}{n}\right)\frac{e^\pi}{Ж_r} \boxplus\right)$$

Where $Ж_1$ = the 1st hyperbolic vortex partition constant, π = Archimedes' constant, e = Euler's number, $Ж_r$ = the hyperbolic vortex radius constant, and e^π = the volume sum of all even-dimensional hyperspheres.

$$\lim_{n\to\infty} \frac{\pi^0}{\Gamma(1)} + \frac{\pi^1}{\Gamma(2)} + \frac{\pi^2}{\Gamma(3)} + \cdots + \frac{\pi^n}{\Gamma(n+1)} = e^\pi$$

$E_h = 4.359744722203833 \ldots \times 10^{-18} \; m^2 kg/s^2$ predicted
$E_h = 4.3597447222071(85) \times 10^{-18} \; m^2 kg/s^2$ measured

Compton wavelength

$$\lambda_C = 2\pi\left(\frac{l_p\, m_p}{m_e}\right)\left(1 + \left(\frac{b}{2}\right)\alpha_F \boxplus\right)$$

Where α_F = the alpha Feigenbaum constant, and $b = 7$.

$\lambda_C = 2.42631023893941 \ldots \times 10^{-12} \; m$ predicted
$\lambda_C = 2.4263102389(16) \times 10^{-12} \; m$ measured

Bohr magneton

$$\mu_B = \frac{ж_1}{2}\left(\frac{l_p{}^2\, q_p\, m_p}{t_p\, m_e}\right)\left(1 + \left(\frac{2}{3}\right)V_{fe}\,ж_\theta{}^2\,\boxplus\right)$$

$\mu_B = 9.27400999397153\ldots \times 10^{-24}\ m^2C/s$ predicted
$\mu_B = 9.274009994(57) \times 10^{-24}\ m^2C/s$ measured

Nuclear magneton

$$\mu_N = \frac{ж_1}{2}\left(\frac{l_p{}^2\, q_p\, m_p}{t_p\, m_+}\right)\left(1 + \frac{1}{n}\left(\frac{2}{3}\right)V_{fe}\,ж_r{}^2\,\boxplus\right)$$

$\mu_N = 5.05078369897026\ldots \times 10^{-27}\ m^2C/s$ predicted
$\mu_N = 5.050783699(31) \times 10^{-27}\ m^2C/s$ measured

Avogadro constant

$$N_A = \frac{6\,ж_1{}^2}{e^\gamma}\left(\frac{1}{q_p\, m_p}\right)\left(1 - \left(2\,ж_2\,P_{up}\right)^2\,\boxplus\right)$$

Where $ж_1$ and $ж_2$ = the 1st and 2nd hyperbolic vortex partition constants, P_{up} = the universal parabolic constant, e = Euler's number, γ = the Euler-Mascheroni constant, a *mol* combines the dimensions of $q_p\, m_p$, and $\boxplus$ = the terminal boundary of the minimal arena.

$N_A = 6.02214076693260\ldots \times 10^{23}\ 1/mol$ predicted
$N_A = 6.02214076 \times 10^{23}\ 1/mol$ previously defined

Stefan-Boltzmann constant

$$\sigma = \frac{\zeta(2)}{2n}\left(\frac{m_p}{t_p{}^3\, T_p{}^4}\right)\left(1 + \left(\frac{P_{up}}{2\,ж_1}\right)^2\,\boxplus\right)$$

Where $\zeta(s)$ = the Reimann zeta function.

$\sigma = 5.67037441935166\ldots \times 10^{-8}\ kg/s^3K^4$ predicted
$\sigma = 5.670374419 \times 10^{-8}\ kg/s^3K^4$ previously defined

molar gas constant

$$R = \frac{6\,\text{ж}_1{}^2}{e^\gamma}\left(\frac{l_p{}^2}{t_p{}^2\,q_p\,T_p}\right)\left(1 - \sqrt{m_R}\,\,Im(\text{ж}_3)^4\,\boxplus\right)$$

Where γ = the Euler-Mascheroni constant, $Im(\text{ж}_3)$ = the imaginary part of the 3rd hyperbolic vortex partition constant, and m_R = Rényi's parking constant.

$$m_R = e^{-2\gamma}\int_0^\infty \frac{e^{-2\,\Gamma(0,x)}}{x^2}$$

$R = 8.31446261914764\ldots\ m^2kg/s^2K\ mol$ predicted
$R = 8.314462618\ m^2kg/s^2K\ mol$ previously defined

proton gyromagnetic ratio

$$\gamma_+ = \left(2\,\text{ж}_r{}^2 P_{up}\right)^2\left(\frac{t_p}{q_p\,m_e}\right)\left(1 + \frac{1}{d_3}\left(3\,Im\left(i^{i^{i^{\cdot^{\cdot}}}}\right)\right)\text{ж}_r{}^3\,\boxplus\right)$$

Where ж_r = the hyperbolic vortex radius, P_{up} = the universal parabolic constant, and $Im\left(i^{i^{i^{\cdot^{\cdot}}}}\right)$ the imaginary part of the infinite power tower of the imaginary unit $i = \sqrt{-1}$.

$\gamma_+ = 2.67522187383442\ldots \times 10^8\ s/C\ kg$ predicted
$\gamma_+ = 2.6752218744(11) \times 10^8\ s/C\ kg$ measured

atomic mass constant

$$m_u = \left(\frac{1}{2n}\right)^3\frac{e^\gamma}{6\,\text{ж}_1{}^2}\left(q_p\,m_p\right)\left(1 + d_2{}^2\left(2\,Re\left(i^{i^{i^{\cdot^{\cdot}}}}\right)\right)\boxplus\right)$$

Where e = Euler's number, γ = the Euler-Mascheroni constant, and $Re\left(i^{i^{i^{\cdot^{\cdot}}}}\right)$ = the Real part of the infinite power tower of i.

$m_u = 1.66053906637533\ldots \times 10^{-27}\ kg$ predicted
$m_u = 1.66053906660(50) \times 10^{-27}\ kg$ measured

electron Thomson cross section

$$\sigma_e = 4\pi\,\text{Ж}_1{}^4\left(\frac{2}{3}\right)\left(\frac{l_p\,m_p}{m_e}\right)^2\left(1 + sin\left(\frac{b\pi}{2\,\Gamma(n)}\right)\text{Ж}_2{}^3\ \boxplus\ \right)$$

Where $n = 5$, $b = 7$, $sin(x) =$ the sine function, $\Gamma(s) =$ the gamma function, and $\text{Ж}_2 =$ the 2nd hyperbolic vortex partition constant.

$\sigma_e = 6.65246160010799\ldots \times 10^{-29}\ m^2$ predicted
$\sigma_e = 6.6524616(18) \times 10^{-29}\ m^2$ measured

Rydberg constant

$$R_\infty = \frac{\text{Ж}_1{}^4}{4\pi}\left(\frac{m_e}{l_p\,m_p}\right)\left(1 - \left(\frac{2\,\Gamma(n)}{d_0}\right)\text{Ж}_\theta{}^3\ \boxplus\ \right)$$

Where $d_0 =$ the 44, and $\text{Ж}_\theta =$ the hyperbolic vortex radian constant.

$R_\infty = 1.09737315685250\ldots \times 10^7\ 1/m$ predicted
$R_\infty = 1.0973731568539(55) \times 10^7\ 1/m$ measured

muon g-factor

$$g_\mu = -\sqrt{b}\,C_{R1}{}^2\,\text{Ж}_1\left(\frac{m_N}{m_\mu}\right)\left(1 + sec\left(cot\left(\frac{2\,\Gamma(n)}{b}\right)\right)\right)\boxplus\ \right)$$

Where $C_{R1} = \cfrac{1}{1+\cfrac{e^{-2\pi}}{1+\cfrac{e^{-4\pi}}{1+\cfrac{e^{-6\pi}}{1+\cdots}}}}$ is Ramanujan's first continued fraction constant.

$g_\mu = -2.00233184124323\ldots$ predicted
$g_\mu = -2.00233184122(82)$ measured

electron g-factor

$$g_e = -\sqrt{b}\,C_{R1}{}^2\,\text{Ж}_1\left(\frac{m_N}{m_\mu}\right)\left(1 - sec\left(cot\left(\frac{2n}{d_0}\right)\right)\text{Ж}_\theta{}^3\ \boxplus\ \right)$$

$g_e = -2.00231930436319\ldots$ predicted
$g_e = -2.00231930436256(35)$ measured

proton g-factor

$$g_+ = 3\sqrt{3\,G_g}\left(\frac{m_+}{m_N}\right)\left(1 + e^{-1/e}\,ж_r{}^3\,\boxplus\right)$$

$g_+ = 5.58569468885333\,\ldots$ predicted
$g_+ = 5.5856946893(16)$ measured

Neutron magnetic moment

$$N_\mu = -\frac{\sqrt{2}\,G_g}{ж_r{}^2}\left(\frac{l_p{}^2 q_p\,m_p}{t_p\,m_+}\right)\left(1 - \left(\frac{4\pi}{17}\right)\boxplus\right)$$

Where G_g = the tether length for grazing half the unit circle, and π = Archimedes' constant.

$N_\mu = -9.66236470868258\,\ldots \times 10^{-27}\,m^2 C/s$ predicted
$N_\mu = -9.6623647(23) \times 10^{-27}\,m^2 C/s$ measured

Wien entropy constant

$$b_{entropy} = L_{LL}{}^{\frac{1}{4}}\left(\frac{b}{n}\right)^{\frac{1}{2}}\left(l_p\,T_p\right)\left(1 - \frac{1}{\sqrt{2}}\boxplus\right)$$

Where $b = 7$, $n = 5$, and L_{LL} = the Laplace limit.

$b_{entropy} = 3.002916077148106\,\ldots \times 10^{-3}\,mK$ predicted
$b_{entropy} = 3.002916077 \times 10^{-3}\,mK$ measured

neutron radius

$$\frac{r_N}{r_e} = \left(\frac{4\pi}{\Gamma(n)}\right)^2$$

Where π = Archimedes' constant, $n = 5$, r_N = the neutron radius, r_e = the classical electron radius, and $\Gamma(s)$ = the gamma function.

$r_N = 7.72554339837953\,\ldots \times 10^{-16}\,m$ predicted
$r_N = 8.0(10) \times 10^{-16}\,m$ measured

proton radius

$$\frac{r_N}{r_+} = -\zeta'(0)$$

Where r_N = the neutron radius, r_+ = the proton radius, and $\zeta'(s)$ = the 1st derivative of the Reimann zeta function.

$r_+ = 8.40702954466144 \ldots \times 10^{-16}\ m$ predicted
$r_+ = 8.414(19) \times 10^{-16}\ m$ measured

The inner gears of persistence are no longer invisible to our gaze.

The simplest constructable theory of balanced derangements, internally based on the minimal geometry of the hyperbolic figure eight knot, defines an arena projected under 5 perpetual actions (time, space, charge, mass and temperature) with Planck constant boundaries. The external *charge* and *mass* boundaries of that arena partition into the exact charge and mass values that define the fundamental particles of matter. And the 44 derangements of that 5-dimensional arena define the constants of Nature.

The 4 forces of Nature (the strong force, the weak force, the electromagnetic force, and gravity) define the closed anti-symmetric connections between the 5 partition boundaries of the minimal arena.

The most external force, prescribed by the gravitational constant, defines the trivial hyperbolic factorization of those 5 boundaries.

$$G = \left(\frac{l_p^{\ 3}}{t_p^{\ 2}\, m_p} \right)\left(1 - \left(\frac{1}{2}\pi\, \text{ж}_r^{\ 2} \right)\Gamma(n)\ \boxplus \right)$$

Where ж_r = the hyperbolic vortex radius, $n = 5$, and $\Gamma(s)$ = the gamma function.

The relationship between the strength of the electromagnetic force and the strong nuclear force (defining the fine-structure constant) is the 1st hyperbolic vortex partition constant squared.

$$\frac{\text{electromagnetic force}}{\text{strong force}} = \alpha = \text{ж}_1^{\ 2}$$

Squaring that ratio again gives the ratio of the classical electron radius compared to the Bohr electron radius.

$$\frac{r_e}{a_0} = Ж_1{}^4$$

The radius of the Helium atom, the Lithium atom, and so on are

$$He^+ = \frac{a_0}{2} \qquad\qquad Li^{2+} = \frac{a_0}{3} \qquad\qquad \ldots$$

And the nonrelativistic ground state wave function of the hydrogen atom is

$$\psi(r) = \frac{a_0{}^{-\frac{3}{2}}}{\Gamma\left(\frac{1}{2}\right)}\, e^{-\frac{r}{a_0}}$$

Where e = Euler's number, a_0 = the Bohr electron radius, and $\Gamma(s)$ = the gamma function.

The division parameters of the minimal arena define the structural parameters of physics.

constants of Nature

$\alpha = 7.29735257295522 \ldots \times 10^{-3}$ — fine-structure constant

$e = 1.60217657405973 \ldots \times 10^{-19} \ C$ — electron charge

$\alpha_G = 1.75182147492404 \ldots \times 10^{-45}$ — gravitational coupling constant

$\mu_0 = 1.25663706140197 \ldots \times 10^{-6} \ m \ kg/C^2$ — magnetic constant

$S_{mi} = 4.41899541452692 \ldots \times 10^{9} \ kg/s \ C$ — Schwinger magnetic induction

$R_K = 2.5812807449400712 \ldots \times 10^{4} \ m^2 kg/s \ C^2$ — von Klitzing constant

$H_C = 3.87404586641820 \ldots \times 10^{-5} \ sC^2/m^2 kg$ — quantized Hall conductance

$\Phi_0 = 2.06783384793306 \ldots \times 10^{-15} \ m^2 kg/s \ C$ — magnetic flux constant

$K_J = 4.8359784854056 \ldots \times 10^{14} \ s \ C/m^2 kg$ — Josephson constant

$q_c = 3.63694755140949 \ldots \times 10^{-4} \ m^2/s$ — quantum of circulation

$G_0 = 7.74809172907834 \ldots \times 10^{-5} \ s \ C^2/m^2 kg$ — conductance quantum

$c_2 = 1.43877687654906 \ldots \times 10^{-2} \ m \ K$ — 2nd radiation constant

$\hbar = 1.05457172603011 \ldots \times 10^{-34} \ m^2 kg/s$ — Planck's constant

$\kappa = 8.98755179228829 \ldots \times 10^{9} \ m^3 kg/s^2 C^2$ — Coulomb's constant

$\varepsilon_0 = 8.85418781277322 \ldots \times 10^{-12} \ s^2 C^2/m^3 kg$ — electric constant

$c = 2.99792458354727 \ldots \times 10^{8} \ m/s$ — speed of light

$F = 9.64853321242908 \ldots \times 10^{4} \ C/mol$ — Faraday constant

$g_N = -3.82608560204984 \ldots$ — neutron g-factor

$G = 6.67384038951738 \ldots \times 10^{-11} \ m^3/s^2 kg$ — gravitational constant

$k_B = 1.38064931695409 \ldots \times 10^{-23} \ m^2 kg/s^2 K$ — Boltzmann constant

$\omega_c = 7.7634409944556 \ldots \times 10^{20} \ 1/s$ — Compton angular frequency

$c_{1L} = 1.19104287769811 \ldots \times 10^{-16} \ m^4 kg/s^3$ — spectral radiance

$c_1 = 3.74177185197781 \ldots \times 10^{-16} \ m^4 kg/s^3$ — 1st radiation constant

$Z_0 = 3.76730313666885 \ldots \times 10^{2} \ m^2 kg/s \ C^2$ — characteristic impedance

$a_0 = 5.29177210889649 \ldots \times 10^{-11} \ m$ — Bohr electron radius

$r_e = 2.81794032505462 \ldots \times 10^{-15} \ m$ — classical electron radius

$E_h = 4.359744722203833 \ldots \times 10^{-18} \ m^2 kg/s^2$ — Hartree energy

$\lambda_C = 2.42631023893941 \ldots \times 10^{-12} \ m$ — Compton wavelength

$\mu_B = 9.27400999397153 \ldots \times 10^{-24} \ m^2 C/s$ — Bohr magneton

$\mu_N = 5.05078369897026 \ldots \times 10^{-27} \ m^2 C/s$ — Nuclear magneton

$N_A = 6.02214076693260 \ldots \times 10^{23} \ 1/mol$ — Avogadro constant

$\sigma = 5.67037441935166 \ldots \times 10^{-8} \ kg/s^3 K^4$ — Stefan-Boltzmann constant

$R = 8.31446261914764 \ldots \ m^2 kg/s^2 K \ mol$ — molar gas constant

$\gamma_+ = 2.67522187383442 \ldots \times 10^{8} \ s/C \ kg$ — proton gyromagnetic ratio

$m_u = 1.66053906637533 \ldots \times 10^{-27} \ kg$ — atomic mass constant

$\sigma_e = 6.65246160010799 \ldots \times 10^{-29} \ m^2$ — electron Thomson x section

$R_\infty = 1.09737315685250 \ldots \times 10^{7} \ 1/m$ — Rydberg constant

$g_\mu = -2.00233184124323 \ldots$ — muon g-factor

$g_e = -2.00231930436319 \ldots$ — electron g-factor

$g_+ = 5.58569468885333 \ldots$ — proton g-factor

$N_\mu = -9.66236470868258 \ldots \times 10^{-27} \; m^2 C/s$ neutron magnetic moment

$b_{entropy} = 3.002916077148106 \ldots \times 10^{-3} \; mK$ Wien entropy constant

$r_N = 7.72554339837953 \ldots \times 10^{-16} \; m$ neutron radius

$r_+ = 8.40702954466144 \ldots \times 10^{-16} \; m$ proton radius

Where the black digits represent previously known values (either measured or geometrically known), and green digits represent extended predictions.

Chapter 10: Euler's number

So far, we have discovered that the simplest constructable theory of balanced derangements, internally based on the minimal geometry of the hyperbolic figure eight knot, defines an arena projected under 5 perpetual actions (time, space, charge, mass and temperature) with Planck constant boundaries. The external *charge* and *mass* boundaries of that arena partition into the exact charge and mass values that define the fundamental particles of matter. And the 44 derangements of this minimal balance of boundaries define the constants of Nature. In other words, the partition parameters of the simplest possible self-balanced geometry precisely define the constructive parameters of physical reality (the parameters of quantum field theory and general relativity).

In this chapter, we notice that the union of the minimal arena's boundaries is *dually* ideally hyperbolic. That is, the Planck union $(U(Planck))$ defined as the sum of the minimal arena's derangements $\sum d_k$ divided by the product of its rotations $\prod \phi_k$, is equal to the binomial factorized union of the ideal hyperbolic connection (represented by Euler's number e) and the ideal hyperbolic partitioning (represented by the gamma function $\Gamma(s)$).

$$U(Planck) = \frac{\sum d_k}{\prod \phi_k} = e \left(1 + \Gamma(s) \boxplus \right)$$

Where

$$\sum_{k=0}^{4} d_k = d_0 + d_1 + d_2 + d_3 + d_4 = 137$$

$$\prod_{k=0}^{4} \phi_k = \phi_0 \phi_1 \phi_2 \phi_3 \phi_4 = 50.3938448477126 \ldots$$

$d_0 = 44$	$\phi_0 = 5.39125836832313 \ldots$
$d_1 = 35$	$\phi_1 = 1.61625918175645 \ldots$
$d_2 = 18$	$\phi_2 = 1.87554596713962 \ldots$
$d_3 = 8$	$\phi_3 = 2.17642683817579 \ldots$
$d_4 = 32$	$\phi_4 = 1.41678698590795 \ldots$

and the argument of the gamma function divides the total derangements of this geometry ($!n$) by the break in scale symmetry of the system

(**b**) and the external condition of split symmetry, represented by $\bar{s}_{lse}$ = the mean line-between-square edges length (in hypercube line picking), defined as follows.[5]

$$\bar{s}_{lse} = \frac{2}{3} \int_0^1 \int_0^1 \sqrt{x^2 + y^2}\, dx\, dy + \frac{1}{3} \int_0^1 \int_0^1 \sqrt{1 + (y - z)^2}\, dz\, dy$$

$$-\bar{s}_{lse} = -\frac{1}{3}\left(\frac{1}{3} P_{up} + \frac{2}{3}\left(1 + \left(\frac{P_{up} - \sqrt{2}}{2} \right) \right) \right)$$

$$P_{up} = \sqrt{2} + sinh^{-1}(1)$$

$$sinh^{-1}(1) = cosh^{-1}(2) = log\left(1 + \sqrt{2} \right)$$

Where P_{up} = the universal parabolic constant, $sinh(x)$ = the hyperbolic sine function, $cosh(x)$ = the hyperbolic cosine function, $log(x)$ = the hyperbolic logarithm function.

---

the Planck union

$$U(Planck) = \frac{\sum d_k}{\prod \phi_k} = e \left(1 + \Gamma \left(\frac{!\,n}{b\, \bar{s}_{lse}} \right) \boxplus \right)$$

Where $\sum d_k$ = the sum of the minimal arena's derangements, $\prod \phi_k$ = the product of its unique rotations, e = Euler's number, $\Gamma(s)$ = the gamma function, $!\,n = 44$ the number of derangements available to 5 rotations, $b = 7$ the break in scale symmetry of this balance, $\bar{s}_{lse}$ = the mean line-between-square edges length, and $\boxplus$ = the terminal boundary of the minimal arena.

---

[5] The total derangements $\sum d_k = 137 = (4^2 + 11^2)$ equals the squared sum of the unique factors in the base derangement $!\,n = 44 = (4 \times 11)$.

80

Euler's number $e = 2.71828182845904\ldots$ defines the base of the hyperbolic logarithm, commonly known as the *natural* logarithm because it is *the* number that hyperbolically generalizes. That is, e defines the infinite sum of inverse factorizations, while its multiplicative inverse defines the *alternating* infinite sum of inverse factorizations.

$$e = \frac{1}{0!} + \frac{1}{1!} + \frac{1}{2!} + \frac{1}{3!} + \frac{1}{4!} + \frac{1}{5!} + \cdots = \sum_{k=0}^{\infty} \frac{1}{k!}$$

$$\frac{1}{e} = \frac{1}{0!} - \frac{1}{1!} + \frac{1}{2!} - \frac{1}{3!} + \frac{1}{4!} - \frac{1}{5!} + \cdots = \sum_{k=0}^{\infty} \frac{(-1)^k}{k!}$$

By generalizing this base exponentiation to any factor x,

$$e^x = \frac{x^0}{0!} + \frac{x^1}{1!} + \frac{x^2}{2!} + \frac{x^3}{3!} + \frac{x^4}{4!} + \frac{x^5}{5!} + \cdots = \sum_{k=0}^{\infty} \frac{x^k}{k!}$$

and splitting that generalization into its even and odd parts, we get the *hyperbolic identities*.

$$\cosh(x) = \frac{x^0}{0!} + \frac{x^2}{2!} + \frac{x^4}{4!} + \frac{x^6}{6!} + \frac{x^8}{8!} + \cdots \qquad \cosh(x) = \frac{1}{2}(e^x + e^{-x})$$

$$\sinh(x) = \frac{x^1}{1!} + \frac{x^3}{3!} + \frac{x^5}{5!} + \frac{x^7}{7!} + \frac{x^9}{9!} + \cdots \qquad \sinh(x) = \frac{1}{2}(e^x - e^{-x})$$

And by alternating the signs of those sequences we get the *circular identities*.

$$\cos(x) = \frac{x^0}{0!} - \frac{x^2}{2!} + \frac{x^4}{4!} - \frac{x^6}{6!} + \frac{x^8}{8!} - \cdots \qquad \cos(x) = \frac{1}{2}\left(e^{xi} + e^{-xi}\right)$$

$$\sin(x) = \frac{x^1}{1!} - \frac{x^3}{3!} + \frac{x^5}{5!} - \frac{x^7}{7!} + \frac{x^9}{9!} - \cdots \qquad \sin(x) = \frac{i}{2}\left(e^{xi} - e^{-xi}\right)$$

The hyperbolic cosine $cosh(x)$ defines the average of the exponential e^x and inverse exponential e^{-x} functions.

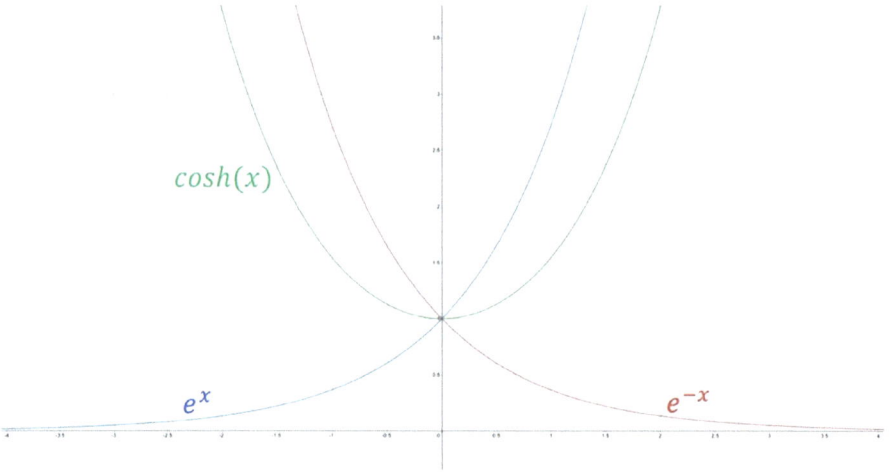

The hyperbolic base of logarithms and the hyperbolic cosine.

These hyperbolic functions have a built-in *golden ratio* dependence. The hyperbolic cosine and the hyperbolic sine squared intersect with coordinates $\left(\mp \log\left(\varphi - \sqrt{\varphi} \right), \varphi \right)$.

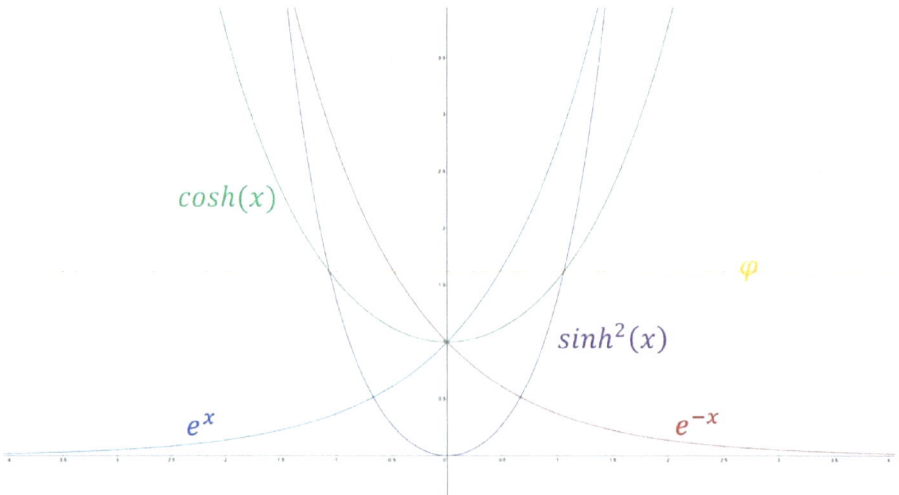

Where φ = the golden ratio, e = Euler's number, $sinh(x)$ = the hyperbolic sine, $cosh(x)$ = the hyperbolic cosine, and $\log(x)$ = the hyperbolic logarithm function.

Under universal binomial factorization the external domain of the minimal arena ideally hyperbolically connects (prescribed by Euler's number e) and ideally hyperbolically partitions (prescribed by the gamma function $\Gamma(s)$).

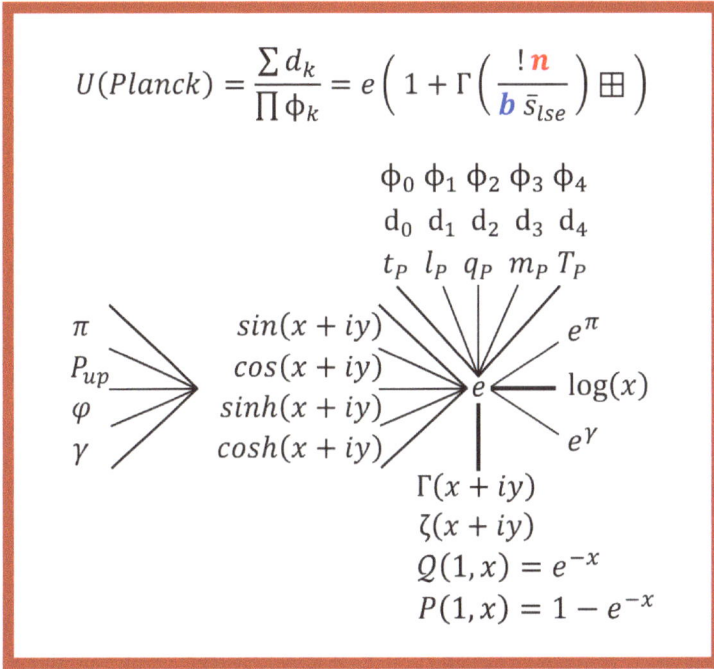

$$U(Planck) = \frac{\sum d_k}{\prod \phi_k} = e\left(1 + \Gamma\left(\frac{!\,n}{b\,\bar{s}_{lse}}\right)\boxplus\right)$$

$$\phi_0 \; \phi_1 \; \phi_2 \; \phi_3 \; \phi_4$$
$$d_0 \; d_1 \; d_2 \; d_3 \; d_4$$
$$t_P \; l_P \; q_P \; m_P \; T_P$$

$$\pi \qquad sin(x+iy)$$
$$P_{up} \qquad cos(x+iy) \qquad e^{\pi}$$
$$\varphi \qquad sinh(x+iy) \qquad e \longleftarrow log(x)$$
$$\gamma \qquad cosh(x+iy) \qquad e^{\gamma}$$

$$\Gamma(x+iy)$$
$$\zeta(x+iy)$$
$$Q(1,x) = e^{-x}$$
$$P(1,x) = 1 - e^{-x}$$

Where e = Euler's number, π = Archimedes' constant, P_{up} = the universal parabolic constant, φ = the golden ratio, γ = the Euler-Mascheroni constant, $\bar{s}_{lse}$ = the mean line-between-square edges length, $sin(x)$ = the sine function, $cos(x)$ = the cosine function, $sinh(x)$ = the hyperbolic sine function, $cosh(x)$ = the hyperbolic cosine function, $log(x)$ = the hyperbolic logarithm function, $\Gamma(s)$ = the gamma function, $\zeta(s)$ = the Reimann zeta function, $Q(a,s)$ = the Q-regularized gamma function, $P(a,s)$ = the P-regularized gamma function, s = the generalized complex number $(x+iy)$, $\sum d_k$ = the complete sum of the minimal arena's derangements, $\prod \phi_k$ = the product of its unique rotations, $!\,n = 44$ the number of derangements available to ($n = 5$) rotations, $b = 7$, and $\boxplus$ = the terminal boundary of the minimal arena.

$\pi = 3.14159265358979\,...$ Archimedes' constant
$P_{up} = 2.29558714939263\,...$ universal parabolic constant
$\varphi = 1.61803398874989\,...$ the golden ratio
$\gamma = 0.577215664901532\,...$ Euler-Mascheroni constant

This ideal hyperbolic balance (functionalized by e^x, $\log(x)$, $\sinh(x)$, $\cosh(x)$, $\sin(x)$, $\cos(x)$) is parameterized by 4 geometric numbers: Archimedes' constant "pi", the universal parameter for parabolas, the golden ratio, and the Euler-Mascheroni constant (π, P_{up}, φ, γ).

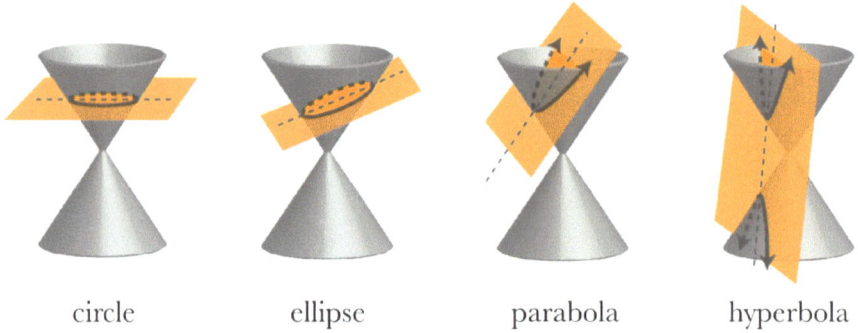

| circle | ellipse | parabola | hyperbola |

Slicing the conic division boundary of the external domain.

Any planar slice taken *perpendicular* to the axis of the external domain's conic division boundary has a *circular* intersection, parametrized by π. At the other extreme, any slice taken *parallel* to one of the external domain's conic division boundaries has a *parabolic* intersection, parametrized by the universal parameter for parabolas (P_{up}).

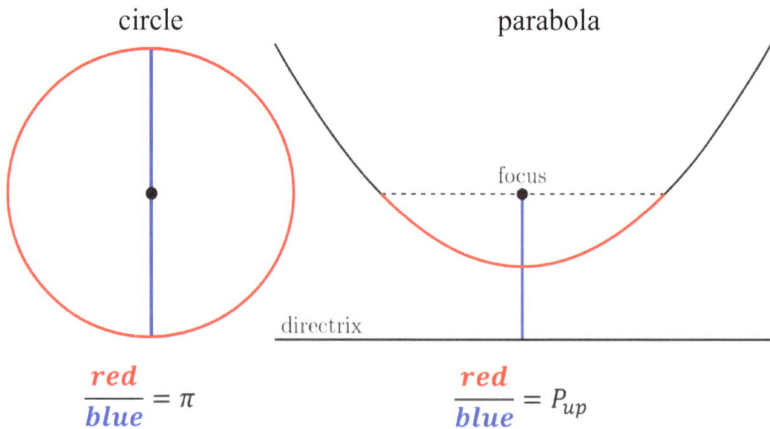

circle   parabola

focus

directrix

$$\frac{\textbf{\textit{red}}}{\textbf{\textit{blue}}} = \pi \qquad\qquad \frac{\textbf{\textit{red}}}{\textbf{\textit{blue}}} = P_{up}$$

Just as the ratio of the circumference of a circle (in red) to its diameter (in blue) is always the universal constant for circles (π), the ratio of the arclength of the parabolic segment formed by the latus rectum of any parabola (in red) and its focal parameter (in blue) is always the universal constant for parabolas (P_{up}).

$$\pi = \int_{-1}^{1} \frac{1}{\sqrt{1-x^2}}\, dx \qquad \pi = \int_{-\infty}^{\infty} \frac{1}{1+x^2}\, dx$$

$$\frac{\pi}{2} = \int_{-1}^{1} \sqrt{1-x^2}\, dx \qquad \sqrt{\pi} = \int_{-\infty}^{\infty} e^{-x^2}\, dx$$

$$P_{up} = \int_{-1}^{1} \sqrt{1+x^2}\, dx \qquad \frac{\pi}{8} = \int_{0}^{1} \sqrt{x(1-x)}\, dx$$

$$2\pi = Ж_1 Ж_2 Ж_3 Ж_4 \qquad \pi = \log(e^{\pi})$$

$$-4\pi = Ж_1{}^2 + Ж_2{}^2 + Ж_3{}^2 + Ж_4{}^2$$

Where π = Archimedes' constant, P_{up} = the universal parabolic constant, $Ж_1, Ж_2, Ж_3, Ж_4$ = the 1st, 2nd, 3rd, and 4th hyperbolic vortex partition constants, e = Euler's number, and $\log(x)$ = the hyperbolic logarithm function.

Shapes that depend on both of these parameters can be found by taking the graph of ($y = e^x$) from $x = -\infty$ to 0, and revolving it around the x-axis, which yields a surface with area = $\pi\, P_{up}$. Or by taking the graph of ($y = \cos(x)$) from $x = -\frac{\pi}{2}$ to $\frac{\pi}{2}$, and revolving it around the x-axis, which yields a surface with area = $2\pi\, P_{up}$.

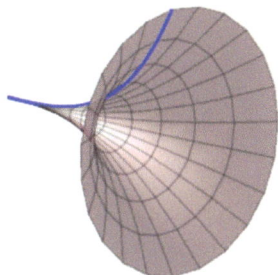

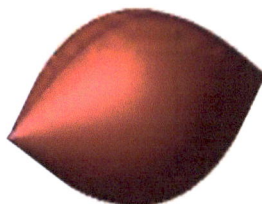

$Area = \pi\, P_{up}$ $Area = 2\pi\, P_{up}$

Surface areas depending on π and P_{up}.

The golden ratio richly defines the unitary recursiveness of the minimal arena's 5-dimensional hyperbolic logarithmic split ($\varphi = \frac{1+\sqrt{n}}{2}$, where $n = 5$).

$$\varphi = \sqrt{1 + \sqrt{1 + \sqrt{1 + \sqrt{1 + \cdots}}}} \qquad \varphi + 1 = \varphi^2$$

$$\sinh(\log(\varphi)) = \frac{1}{2} \qquad \varphi - 1 = \frac{1}{\varphi}$$

$$\varphi = 1 + \cfrac{1}{1 + \cfrac{1}{1 + \cfrac{1}{1 + \cfrac{1}{1 + \cdots}}}} \qquad \frac{1}{\varphi} = 0 + \cfrac{1}{1 + \cfrac{1}{1 + \cfrac{1}{1 + \cfrac{1}{1 + \cdots}}}}$$

Where φ = the golden ratio, $\sinh(x)$ = the hyperbolic sine function, and $\log(x)$ = the hyperbolic logarithm function.

The limiting ratio of successive terms in the Fibonacci sequence, or any Fibonacci-like sequence (Kepler), is the golden ratio.

$$\lim_{n \to \infty} \frac{F_{n+1}}{F_n} = \varphi$$

Where F_n = the nth Fibonacci number.

The entire Fibonacci sequence can be captured in closed-form in terms of the golden ratio.

$$F(n) = \frac{\varphi^n - (-\varphi)^{-n}}{\sqrt{5}}$$

And successive powers of φ obey the Fibonacci recurrence relation.

$$\varphi^{n+1} = \varphi^n + \varphi^{n-1}$$

86

The Euler-Mascheroni constant defines how the logarithm (which defines the area or "quadrature" of a hyperbola) smoothly connects the internal quantization of the minimal arena.

$$\gamma = \int_0^\infty \log\left(\log\left(\frac{1}{x}\right)\right) dx \qquad \gamma = \int_0^1 e^{-x} \log(x)\ dx$$

$$\gamma = \int_0^\infty \frac{e^{-|x|} \log|x|}{2} dx \qquad \gamma = \int_0^\infty \left(\frac{1}{1-e^{-x}} - \frac{1}{x}\right) e^{-x}\ dx$$

$$\gamma = \int_1^\infty \left(-\frac{1}{x} + \frac{1}{\lfloor x \rfloor}\right) dx \qquad \gamma = 2\int_0^\infty \frac{e^{-x^2} - e^{-x}}{x} dx$$

$$\gamma = \sum_{n=2}^\infty (-1)^n \frac{\zeta(n)}{n} \qquad \gamma = \sum_{n=1}^\infty \sum_{k=2^n}^\infty \frac{(-1)^k}{k}$$

$$\gamma = -\Gamma'(1) \qquad \frac{e^{\pi/2} + e^{-\pi/2}}{\pi\ e^\gamma} = \prod_{n=1}^\infty e^{-1/n}\left(1 + \frac{1}{n} + \frac{1}{2n^2}\right)$$

Where π = Archimedes' constant, γ = the Euler-Mascheroni constant, e = Euler's number, $\log(x)$ = the hyperbolic logarithm function, $\Gamma(s)$ = the gamma function, $\zeta(s)$ = the Reimann zeta function, $|x|$ = the absolute value function, and $\lfloor x \rfloor$ = the floor function.

This ideal discreteness parameter is equal to the limiting difference between the harmonic series and the hyperbolic logarithm. In other words, the Euler-Mascheroni constant is equal to the limiting difference between the quantized area of the harmonic series and the smooth area of the graph of ($y = \frac{1}{x}$) on the interval from $x = 1$ to ∞ (the blue area).

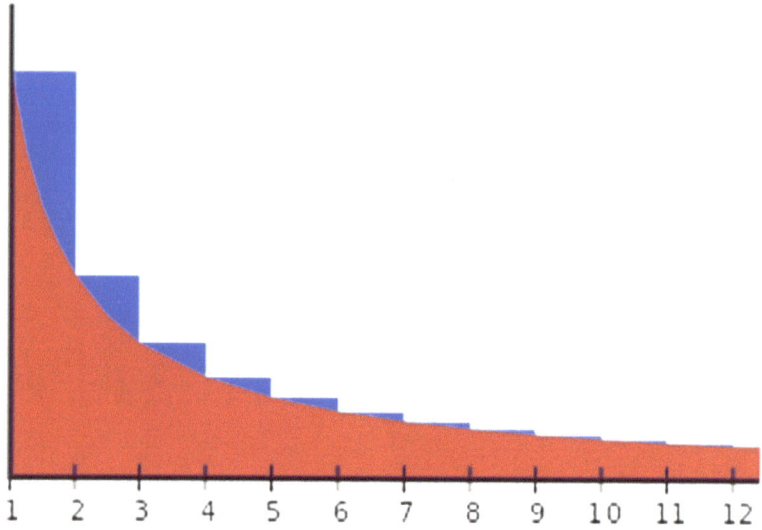

$$\gamma = \textbf{\textit{blue area}} = \lim_{n\to\infty}\left(\sum_{k=1}^{n}\frac{1}{k} - \int_{1}^{n}\frac{1}{x}\,dx\right)$$

Beneath the graph of the hyperbolic logarithm ($y = \frac{1}{x}$) from $x = 0$ to ∞ there is a unit square, with area $= 1$. The next unit area under the curve is bounded by 1 and e, the next by e and e^2, and so on.

$$1 = \int_{1}^{e}\frac{1}{x}\,dx = \int_{e}^{e^2}\frac{1}{x}\,dx = \cdots = \int_{e^n}^{e^{n+1}}\frac{1}{x}\,dx$$

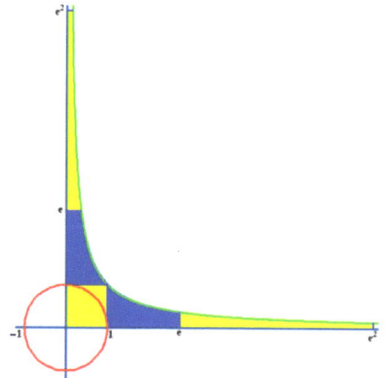

The symmetric unitary square division of the hyperbolic logarithm.

Taking the same graph $\left(y = \frac{1}{x} \right)$ from $x = 1$ to ∞ and revolving it around the x-axis yields a geometric shape known as Gabriel's horn, with an infinite area and a finite volume $= \pi$.

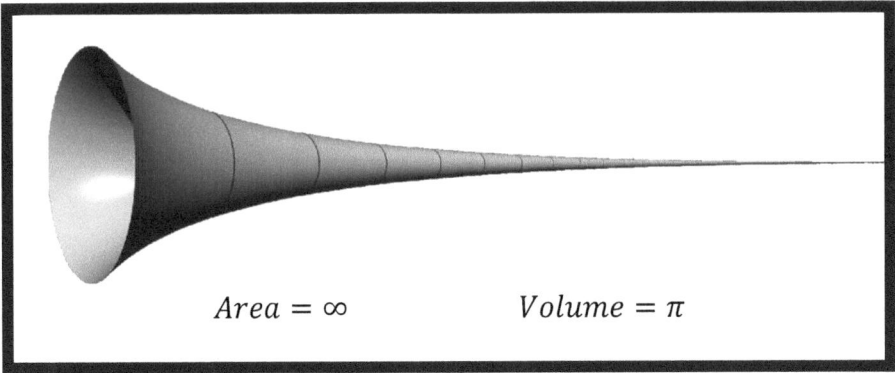

$$Area = \infty \qquad Volume = \pi$$

Now that we've explored the Planck union's ideal hyperbolic connection (characterized by Euler's number e and the four geometric connectors of projective hyperbolic geometry π, P_{up}, φ, and γ), let's take a look at how the Planck union terminally *ideally hyperbolically factors* (represented by the gamma function).

Chapter 11: the gamma function

In this chapter, we make the delightful discovery that the gamma function (the elementary *factorization* function) maps the partition balance of the minimal arena.

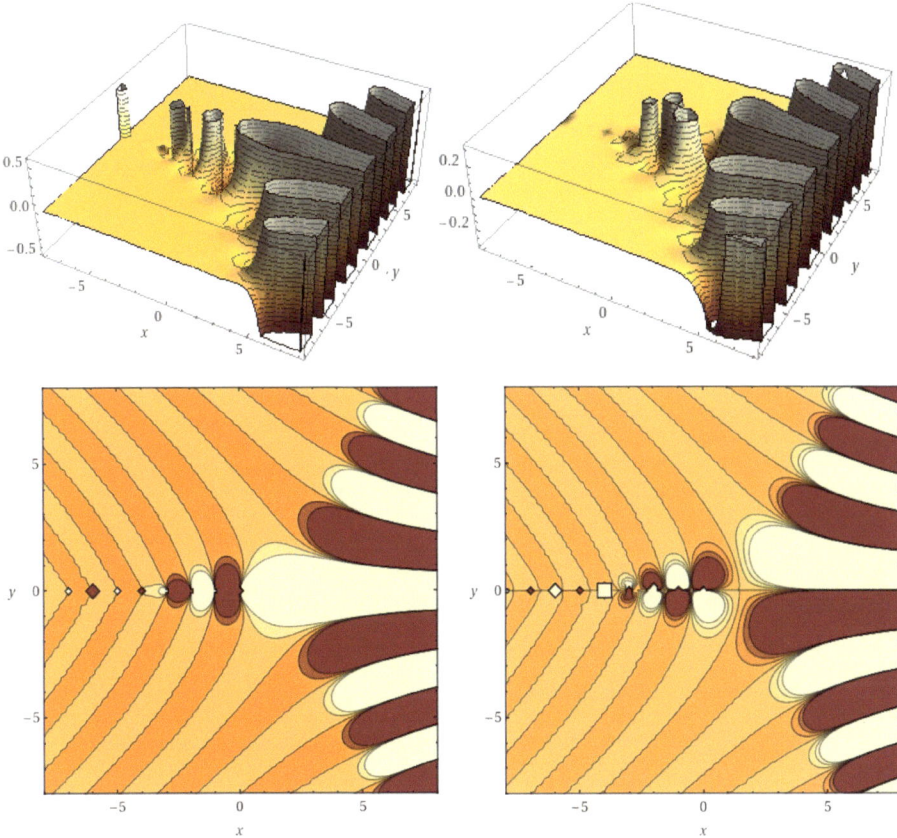

Graph 11: real (left) and imaginary (right) plots of the gamma function under complex argument $\Gamma(x + iy)$.

By symmetry, any successful theory of derangements ($!n$) must be convertible into a theory of factorizations ($n!$).

$$!n = n! \sum_{k=0}^{n} \frac{(-1)^k}{k!}$$

Definition of the derangement function ($!n$) in terms of the factorial function ($n!$).

Therefore, the theory of everything that we just laid out in terms of hyperbolically balanced derangements must also be simultaneously frameable in terms of hyperbolic factorizations, bringing us to the function many luminaries have declared to be "the most beautiful function of all".

The gamma function is the unique analytic continuation of the factorial function. This means that instead of being defined only at positive integer values, like the "primitive" factorial function, the gamma function is defined everywhere, except for its simple poles at $s = 0, -1, -2, -3 \ldots$

$$\Gamma(s) = \int_0^\infty \frac{x^{s-1}}{e^x} \, dx \qquad Re(s) > 0$$

Integral definition of the gamma function. Where $s = (x + iy)$.

In short, the gamma function is the *complete* factorization (partition) function. Let's explore some of its known properties.

The gamma function has no zeros. That is, there is no complex number s for which $\Gamma(s) = 0$. Therefore, the reciprocal gamma function $\frac{1}{\Gamma(s)}$ is an entire function, with zeros at $s = 0, -1, -2, -3 \ldots$

The gamma function hyperbolically generalizes and has a special point, a local minimum x_{min}.

$$\left| \Gamma\left(0 + xi \right) \right|^2 = \frac{\pi}{x \sinh(x\pi)}$$

$$\left| \Gamma\left(\frac{1}{2} + xi \right) \right|^2 = \frac{\pi}{\cosh(x\pi)}$$

$$\left| \Gamma\left(1 + xi \right) \right|^2 = \frac{x\,\pi}{\sinh(x\pi)}$$

$x_{min} = 1.461632144968362 \ldots$ minimum positive argument value of Γ
$\Gamma(x_{min}) = 0.885603194410888 \ldots$ minimal Γ value for positive argument

92

The gamma function is the natural tool for calculating the volume or area of an n-dimensional hypersphere, the arc length of the lemniscate (Chapter 4) and the arc lengths of ellipses, which are given by elliptic integrals simply evaluated in terms of the gamma function.

$$\Gamma(0, x) = -Ei(-x) = Shi(x) - Chi(x)$$

Where $\Gamma(0, x) =$ the incomplete gamma function, $Ei(x) =$ the elliptic integral, $Shi(x) =$ the hyperbolic sine integral, and $Chi(x) =$ the hyperbolic cosine integral.

Algebraically, the gamma function is the unique function that simultaneously satisfies the following 3 conditions for all complex numbers $s = (x + iy)$.

$$\Gamma(1) = 1 \qquad \Gamma(s + 1) = s\,\Gamma(s) \qquad \lim_{n \to \infty} \frac{\Gamma(s + n)}{\Gamma(n)\, n^s} = 1$$

Internally, the gamma function naturally splits up into 2 regularized 2-dimensional parts $Q(a, s)$ and $P(a, s)$ defined by the following symmetries.

regularized gamma-functions

$$Q(a, s) + P(a, s) = 1 \qquad\qquad \text{sum of regularized } \Gamma \text{ functions}$$

$$\frac{d}{ds} Q(a, s) = -\frac{e^{-s}\, s^{a-1}}{\Gamma(a)} \qquad\qquad \text{equally split derivatives}$$

$$\frac{d}{ds} P(a, s) = +\frac{e^{-s}\, s^{a-1}}{\Gamma(a)} \qquad\qquad \text{equally split derivatives}$$

$$\frac{d}{ds} Q(a, s) + \frac{d}{ds} P(a, s) = 0 \qquad\qquad \text{derivative balance}$$

The unitary generalized values of these Q and P regularized gamma functions are equal to the generalized inverse exponentiation function e^{-x} and its geometric reflection $1 - e^{-x}$.

$$Q(1, x) = e^{-x} \qquad\qquad P(1, x) = 1 - e^{-x}$$

93

In other words, the gamma function internally pulls apart into pieces whose unitary generalized actions define the inverse and reflected-inverse actions of the hyperbolic functions. Its parts constructively generalize in opposition and reflected opposition to the hyperbolic base, Euler's number e.

Special known values of the gamma function include:

$$\frac{1}{\Gamma(0)} = \frac{1}{\Gamma(-1)} = \frac{1}{\Gamma(-2)} = \frac{1}{\Gamma(-3)} = \frac{1}{\Gamma(-4)} = \frac{1}{\Gamma(-5)} = \cdots = 0$$

$\Gamma(1) = 1$

$\Gamma(2) = 1$ $\qquad$ $\Gamma\left(\frac{1}{2}\right) = \sqrt{\pi}$ $\qquad$ $\Gamma\left(\frac{1}{2}\right)\Gamma\left(-\frac{1}{2}\right) = -2\pi$

$\Gamma(3) = 2$

$\Gamma(4) = 6$ $\qquad$ $\Gamma\left(-\frac{1}{2}\right) = -2\sqrt{\pi}$ $\qquad$ $\Gamma\left(\frac{3}{2}\right)\Gamma\left(-\frac{3}{2}\right) = \frac{2}{3}\pi$

$\Gamma(n)\,\zeta(2) = (2\pi)^2$ $\qquad$ $\Gamma\left(\frac{3}{2}\right) = \frac{1}{2}\sqrt{\pi}$ $\qquad$ $\Gamma\left(\frac{b}{2}\right) = -\frac{n}{2}\left(\frac{3}{4}\sqrt{\pi}\right)$

$\Gamma(6)\,\zeta(-3) = 1$

$\Gamma(6)\,\zeta(-b) = 2$ $\qquad$ $\Gamma\left(-\frac{3}{2}\right) = \frac{4}{3}\sqrt{\pi}$ $\qquad$ $\Gamma'(-4)\,(\zeta(2))^2 = \frac{1}{2}$

$$F_{FR} = \int_0^\infty \frac{1}{\Gamma(x)}\,dx \qquad\qquad W_{We} = (\,2^{2n}\,e^\pi\,)^{1/8}\,\frac{\Gamma\left(\frac{1}{2}\right)}{\left(\Gamma\left(\left(\frac{1}{2}\right)^2\right)\right)^2}$$

$$\omega_1 = \frac{\Gamma^3\left(\frac{1}{3}\right)}{4\pi}\,(-1)^{\frac{1}{3}} \qquad\qquad C_{PTA} = \frac{1}{\sqrt{2\pi^n}}\left(\Gamma\left(\left(\frac{1}{2}\right)^2\right)\right)^2$$

$$\omega_2 = \frac{\Gamma^3\left(\frac{1}{3}\right)}{4\pi} \qquad L = \frac{1}{2}s \qquad s = \frac{1}{\sqrt{2\pi}}\left(\Gamma\left(\left(\frac{1}{2}\right)^2\right)\right)^2$$

Where π = Archimedes' constant, F_{FR} = the Fransén-Robinson constant, W_{We} = the Weierstrass constant, C_{PTA} = the Pythagorean triple constant for areas, ω_1 = the omega_1 constant, ω_2 = the omega_2 constant, L = the lemniscate constant, s = the arclength of the unitary lemniscate, $\Gamma(s)$ = the gamma function, $\zeta(s)$ = the Reimann zeta function, n = 5, and b = 7.

The gamma function has a built-in unitary symmetric dependence on the golden ratio. That is,

$$\Gamma(x + 1) = \Gamma(x - 1)$$

has solutions

$$x = \frac{1 \pm \sqrt{5}}{2}$$

where

$$\frac{1 + \sqrt{5}}{2} = \varphi \qquad \frac{1 - \sqrt{5}}{2} = -\frac{1}{\varphi}$$

and φ = the golden ratio.

The gamma function also has a rich limiting dependence on the Euler-Mascheroni constant.

$$\lim_{s \to 1^+} \sum_{n=1}^{\infty} \left(\frac{1}{n^s} - \frac{1}{s^n} \right) = \gamma \qquad \lim_{x \to \infty} \left(x - \Gamma\left(\frac{1}{x}\right) \right) = \gamma$$

$$\lim_{x \to 0} \frac{1}{x} \left(\frac{1}{\Gamma(1 + x)} - \frac{1}{\Gamma(1 - x)} \right) = 2\gamma \qquad \lim_{x \to 0} \left(\Gamma(x) - \frac{1}{x} \right) = -\gamma$$

$$\lim_{x \to 0} \frac{1}{x} \left(\frac{1}{\Psi(1 - x)} - \frac{1}{\Psi(1 + x)} \right) = \frac{\pi^2}{3\gamma^2} \qquad \lim_{x \to 0} \left(\Psi(x) + \frac{1}{x} \right) = -\gamma$$

Where γ = the Euler-Mascheroni constant, π = Archimedes' constant, $\Gamma(s)$ = the gamma function, and $\Psi(s)$ = the digamma function.

To unveil the geometry of the gamma function's ideal hyperbolic factorizations, we set the arguments of the Q regularized gamma function to the rotations of the time and space boundaries $Q(\phi_0, \phi_1)$ and examine its output in terms of the universal binomial factorization prescription.

95

What we find is that, the Q regularized gamma function defines a domain under simple division of the minimum 3-manifold (G_{Gi}), terminally maintaining a 3-dimensional hypersphere connection ($3e^{\pi}$) structured over a double cover ($2n$) lattice of that minimum manifold.

Q regularized under the rotations of time and space

$$Q(\phi_0, \phi_1) = \left(\frac{1}{G_{Gi}}\right)\left(1 - 3e^{\pi}\left(\frac{1}{G_{Gi}}\right)^{2n} \boxplus\right)$$

Where $Q(a, s)$ = the Q regularized gamma function, ϕ_0 the rotation of the time boundary, ϕ_1 = the rotation of the space boundary, G_{Gi} = Gieseking's constant, e^{π} = the volume sum of all even dimensional hyperspheres, and $\boxplus$ = the terminal boundary of the minimal arena.

To get the other (orthogonal) half of the story, we set the arguments of the P regularized gamma function to the same time and space rotations $P(\phi_0, \phi_1)$ which, under universal binomial construction, reveal the internal and external partition parameters of the minimal arena.

P regularized under the rotations of time and space

$$P(\phi_0, \phi_1) = \left(\frac{b}{\delta_F{}^4}\right)\left(1 - \varkappa_2\sqrt{\frac{2}{W_{We}}} \boxplus\right)$$

Where $P(a, s)$ = the P regularized gamma function, ϕ_0 the rotation of the time boundary, ϕ_1 = the rotation of the space boundary, $b = 7$ the break in scale symmetry, δ_F = the delta Feigenbaum constant representing the limiting bifurcation velocity, $\varkappa_2$ = the 2nd hyperbolic vortex partition constant, W_{We} = the Weierstrass constant, and $\boxplus$ = the terminal boundary of the minimal arena.

The universal binomial factorization prescription allows us to fully unveil the gamma function's balance of internal and external rotations (factors). Setting the arguments of the Q and P regularized gamma functions to the 2 internal (time and space) rotations of the minimal arena (ϕ_0, ϕ_1), and joining those arguments under ideal bifurcation (square/square-root) arrangement, equals externally setting the gamma function to divide the 4th rotation of the minimal arena by its 2nd and 3rd rotations, terminally bound as the minimum non-compact 3-manifold (G_{Gi}), persisting as a projected balance of n rotations maintained over triple doubly-periodic factorization (W_{We}).

---

the geometry of the gamma function

$$\left(\delta_F\, Q(\phi_0, \phi_1)\right)^2 \sqrt{(P(\phi_0, \phi_1))} = \Gamma\left(\frac{\phi_4}{\phi_2\,\phi_3}\right)\left(1 - (G_{Gi})\frac{n}{W_{We}^3}\boxplus\right)$$

internal partitions   external partitions

Where δ_F = the delta Feigenbaum constant, ϕ_0 the rotation of the time boundary, ϕ_1 = the rotation of the space boundary, ϕ_2 the rotation of the charge boundary, ϕ_3 = the rotation of the mass boundary, ϕ_4 the rotation of the temperature boundary, G_{Gi} = Gieseking's constant, n = 5 the number of unique rotations partitioning the minimal arena, W_{We} = the Weierstrass constant, $Q(a, s)$ = the Q regularized gamma function, $P(a, s)$ = the P regularized gamma function, $\Gamma(s)$ = the gamma function, and $\boxplus$ = the terminal boundary of the minimal arena.

The *most beautiful function of all* defines the dimensions of physical measure (time, space, charge, mass, and temperature) as the mutually balanced boundaries of the minimal arena, and scripts the algebraic structure of persistent form in terms of geometry's simplest 3-manifold.

Under unitary and inverse-complex argument, the Q regularized gamma function $Q(1, (x + iy)^{-1})$ (Graph 12) is the inverse graph of the internal action of balance 1 (Graph 3).

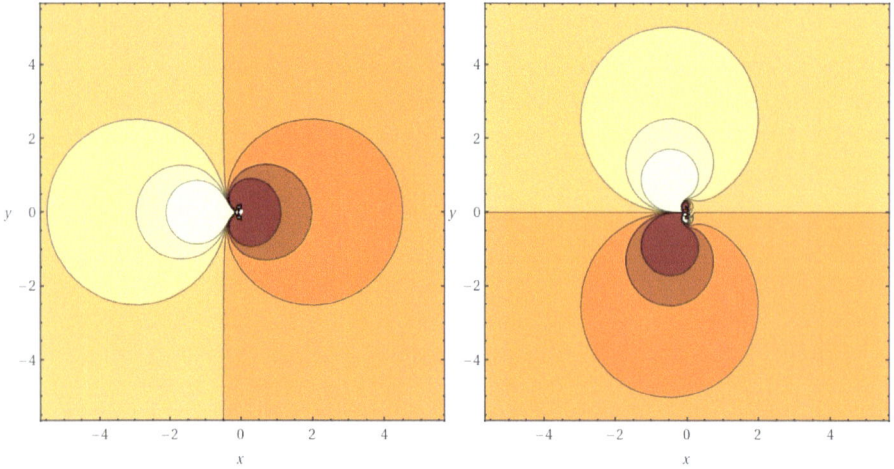

Graph 12: real (left) and imaginary (right) plots of the Q regularized gamma function under unitary and inverse-complex argument. $Q(1, (x + iy)^{-1})$. Compare to balance 1 (Graph 3).

Zooming in on the detail around (0,0) (Graph 13) reproduces the dual inverse-complex action of balance 2 dubbed "the partition butterfly" (Graph 6). Zooming out, reproduces the dual complex action of balance 2 (Graph 5).

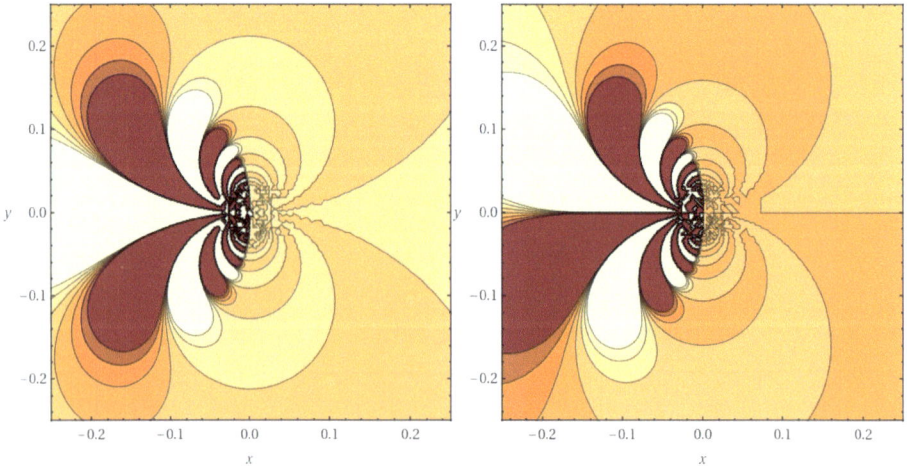

Graph 13: real (left) and imaginary (right) plots of the Q regularized gamma function under unitary and inverse-complex argument, zoomed in. $Q(1, (x + iy)^{-1})$ Compare to balance 2 (Graph 6).

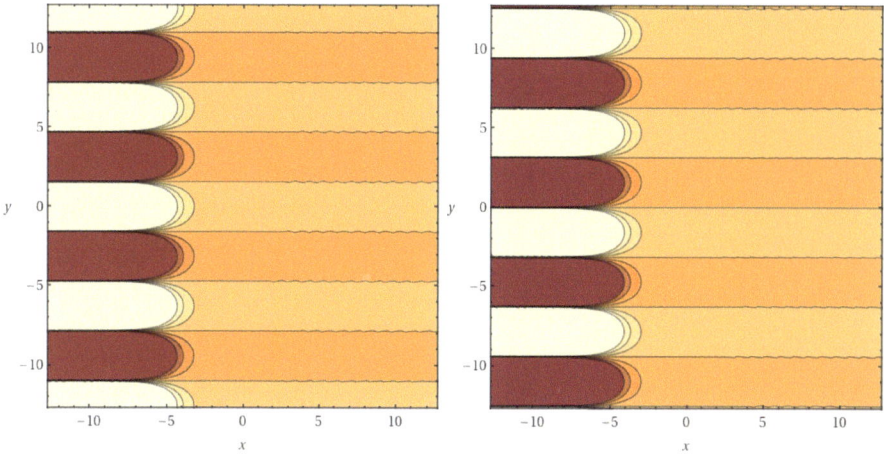

Graph 14: real (left) and imaginary (right) plots of the Q regularized gamma function under unitary and complex argument $Q(1, (x + iy))$. Compare to balance 2 (Graph 5).

And under triple (unitary, complex, and inverse-complex) arguments the P regularized gamma function $P(1, (x + iy), (x + iy)^{-1})$ maps the self-dual external (real) and internal (imaginary) partition space. Note the unit circle in the real factorization map (Graph 15).

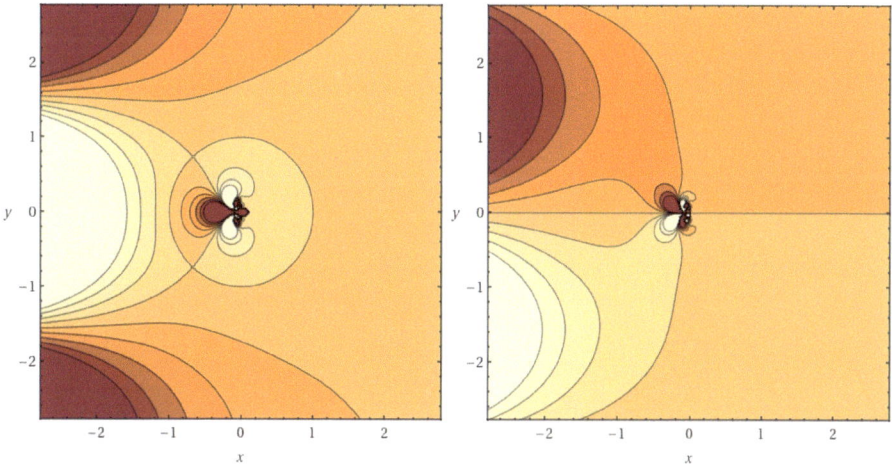

Graph 15: real (left) and imaginary (right) plots of the P regularized gamma function under unitary, complex, and inverse-complex argument $P(1, (x + iy), (x + iy)^{-1})$.

99

An awareness of the gamma function's central role in the persistent balance of reality, coupled with an awareness of the binomial construction of that balance, suddenly equips us with the ability to ask many new precise questions about reality and to discover their geometric solutions. For example, we can now ask, "On what boundaries does the gamma function factor? And what is the geometry of those factorizations?"

To find the gamma function's trivial factorization boundary we set the gamma function's argument to the sum of the minimal arena's 5 rotations. Then we ask, "On what division boundary B_p and with what terminal geometric action A_t does this factorization trivially close (become equal to its own complete inverse)?"

$$\Gamma\left(\sum_{k=0}^{4} \phi_k\right) = \int_0^\infty \frac{1}{\Gamma(x)} dx \left(\frac{1}{B_p}\right)(1 + A_t \boxplus)$$

Where the sum of rotations is,

$$\sum_{k=0}^{4} \phi_k = \phi_0 + \phi_1 + \phi_2 + \phi_3 + \phi_4 = 12.4762773413029 \dots$$

and the gamma function's complete inverse is defined as the integral of the inverse gamma function from zero to infinity, known as the Fransén-Robinson constant (F_{FR}).

$$F_{FR} = \int_0^\infty \frac{1}{\Gamma(x)} dx = 2.80777024202851 \dots$$

Just like before, this equation has 2 unknowns, and since the terminal boundary of this geometry doesn't come into play until the 7th digit of this action, those 2 unknowns can be solved in pieces. That is, since $\boxplus = \left(\frac{l_p \, m_p}{q_p^2}\right) = 1 \times 10^{-7}$ we can safely ignore the geometric contribution of the terminal factor at first, solve for its primary boundary B_p, and then solve for its terminal geometric action A_t.

By temporarily setting $A_t = 0$ and solving for B_p (expecting an answer that is correct to approximately the first 7 digits) we get:

$$\Gamma\left(\sum_{k=0}^{4}\phi_k\right) = \int_0^\infty \frac{1}{\Gamma(x)}\,dx\left(\frac{1}{B_p}\right) \quad B_p = 2.17636464329325\ldots\times 10^{-8}$$

$$m_p = 2.17642683817579\ldots\times 10^{-8}$$

Which identifies the Planck mass boundary $B_p = m_p$.

Setting $B_p = m_p$, we then solve for the terminal geometric action A_t of the gamma function's trivial partition balance: $A_t = \left(\frac{d_0}{n}\right)\sqrt{L}\;ж_r{}^2$.

$$\Gamma\left(\sum_{k=0}^{4}\phi_k\right) = \int_0^\infty \frac{1}{\Gamma(x)}\,dx\left(\frac{1}{m_p}\right)\left(1 + \left(\frac{d_0}{n}\right)\sqrt{L}\;ж_r{}^2\;\boxplus\right)$$

Where $\Gamma(s) =$ the gamma function, $\sum\phi_k =$ the sum of the minimal arena's unique rotations, $m_p =$ the Planck mass, $ж_r =$ the hyperbolic vortex radius constant, $L =$ the lemniscate constant, and $d_0 = !\,n =$ the full 44 derangements available to ($n = 5$) unique partitions.

To inquire about the geometry of the gamma function's other 2 external factorization boundaries (q_P and T_P) we follow the same procedure for the *squared* and *fourth* powered gamma factorizations of the same rotations. That is, under universal binomial factorization we set the division boundaries of the squared and fourth power gamma factorizations equal to the Planck charge and the inverse Planck temperature boundaries. Then for each we ask, "What is the primary action of this factorization, and what is its terminal action?"

$$\Gamma^2\left(\sum_{k=0}^{4}\phi_k\right) = A_{p_2}\left(\frac{1}{q_p}\right)\left(1 + A_{t_2}\;\boxplus\right)$$

$$\Gamma^4\left(\sum_{k=0}^{4}\phi_k\right) = A_{p_4}\left(T_p\right)\left(1 + A_{t_4}\;\boxplus\right)$$

Where $q_p =$ the Planck charge boundary, and $T_p =$ the Planck temperature boundary, $\Gamma(s) =$ the gamma function, $\sum\phi_k =$ the sum of the minimal arena's unique rotations, and $\boxplus =$ the terminal boundary of the minimal arena.

Solving first for the primary action and then the terminal action of each, we find that the unique rotations of the minimal arena factor about the Planck mass, charge and temperature boundaries as follows.

factorization map

$$\Gamma\left(\sum_{k=0}^{4}\phi_k\right) = F_{FR}\left(\frac{1}{m_p}\right)\left(1 + \left(\frac{d_0}{n}\right)\sqrt{L}\,\text{ж}_r{}^2\,\boxplus\right)$$

$$\Gamma^2\left(\sum_{k=0}^{4}\phi_k\right) = \left(\frac{d_3}{d_4}\right)\frac{\lambda_{GD}}{n}\left(\frac{1}{q_p}\right)\left(1 + 3\left(\frac{d_1}{d_0}\right)C_{PTA}{}^4\,\text{ж}_r{}^4\,\boxplus\right)$$

$$\Gamma^4\left(\sum_{k=0}^{4}\phi_k\right) = \left(\frac{d_2}{d_3}\right)\bar{s}_{lse}\left(T_p\right)\left(1 + \left(\frac{n}{\Gamma(n)}\right)F_{FF}{}^4\,\text{ж}_r{}^4\,\boxplus\right)$$

Where m_p = the Planck mass boundary, q_p = the Planck charge boundary, T_p = the Planck temperature boundary, d_k = the number of derangements participating in the kth balance of the minimal arena, $n = 5$, φ = the golden ratio, ж_r = the hyperbolic vortex radius constant, L = the lemniscate constant, F_{FR} = the Fransén-Robinson constant, λ_{GD} = the Golomb-Dickman constant, $\bar{s}_{lse}$ = the mean line-between-square edges length, C_{PTA} = the Pythagorean triple constant for areas, F_{FF} = the Fibonacci factorial constant, $li(x)$ = the logarithmic integral, $\Gamma(s)$ = the gamma function, $\sum\phi_k$ = the sum of the minimal arena's unique rotations, and $\boxplus$ = the terminal boundary of the minimal arena.

$$\lambda_{GD} = \int_0^1 e^{li(x)}\,dx \qquad C_{PTA} = \frac{1}{\sqrt{2\pi^n}}\left(\Gamma\left(\left(\frac{1}{2}\right)^2\right)\right)^2$$

$$F_{FR} = \int_0^\infty \frac{1}{\Gamma(x)}\,dx \qquad F_{FF} = \prod_{k=1}^\infty\left(1 - \left(-\frac{1}{\varphi}\right)^k\right)$$

$$\bar{s}_{lse} = \frac{2}{3}\int_0^1\int_0^1 \sqrt{x^2 + y^2}\,dx\,dy + \frac{1}{3}\int_0^1\int_0^1 \sqrt{1 + (y - z)^2}\,dz\,dy$$

The universal binomial factorization prescription is a powerful new tool for waking ourselves up to the features of existence we've remained blind to. Under that prescription, every clear question we ask the gamma function makes us aware of a unique geometric feature of persistence, enabling us to see features of reality that have always been there, but remained hidden from our conscious view (like seeing the number of an object's sides was hidden from view before Euler).

For example, if we ask, "How does the time boundary factor ($\Gamma(\phi_0)$)?" the universal binomial factorization prescription tells us that it factors into seven circles ($2\pi b$) terminally arranged into a split squared balanced derangement of the minimum manifold.

$$\Gamma(\phi_0) = 2\pi \, b \left(1 + \frac{1}{b} \left(\frac{n^2}{\sqrt{d_0}} \right) \left(\frac{Ж_r^2}{G_{Gi}} \right)^2 \boxplus \right)$$

Where π = Archimedes' constant, $Ж_r$ = the hyperbolic vortex radius constant, G_{Gi} = Gieseking's constant, $\Gamma(s)$ = the gamma function, $n = 5$ the number of unique rotations partitioning the minimal arena, $b = 7$ the break in scale symmetry maintained between the 2 internal boundaries of that arena, $d_0 = !\,n$, where $!\,n = $ the derangement function, and $\boxplus$ = the minimal arena's terminal boundary.

The universal binomial factorization prescription also empowers us to explore the gamma function's sister, the Riemann zeta function.

Chapter 12: the Riemann zeta function

The Riemann hypothesis, widely regarded the most important unsolved problem in pure mathematics, conjectures that the zeros of the Riemann zeta function occur only at negative even integers, and at positive complex values $s = (x + iy)$ with real part $x = \frac{1}{2}$.[6]

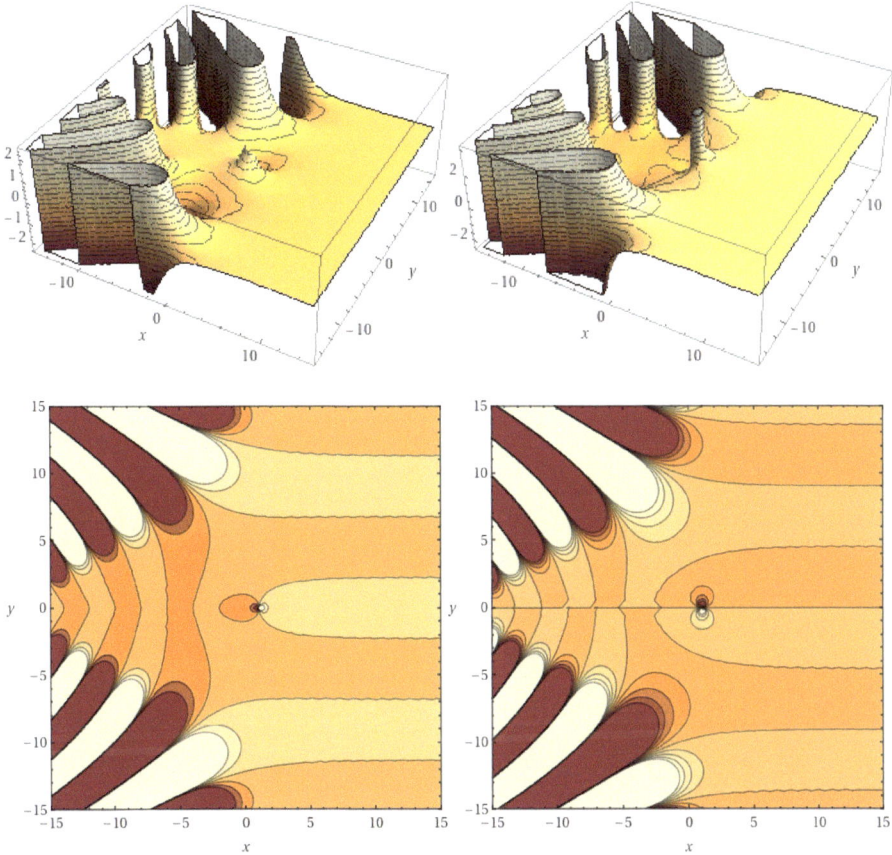

Graph 16: real (left) and imaginary (right) plots of the Riemann zeta function under complex argument $\zeta(x + iy)$.

---

[6] The zeros of the Riemann zeta function intrinsically relate to the distribution of prime numbers.

The Riemann zeta function is defined as an infinite sum *and* as an infinite product over the primes.

$$\zeta(s) = \sum_{n=1}^{\infty} \frac{1}{n^s} \qquad\qquad \zeta(s) = \prod_{p} \frac{1}{1 - p^{-s}}$$

Where $s = (x + iy)$ is a complex number, and p is a prime number.

The Reimann zeta function and the gamma function are intimately connected. When $Re(s) > 1$ the Riemann zeta function can be represented in terms of the inverse gamma function multiplied by the converging summation (integral) form of the gamma function offset by 1.

$$\zeta(s) = \frac{1}{\Gamma(s)} \int_{0}^{\infty} \frac{x^{s-1}}{e^x - 1} dx \qquad\qquad Re(s) > 1$$

bounded by $y \geq 3$, and $x \geq 1 - \dfrac{1}{(\log|y|)^{\frac{2}{3}}(\log \log|y|)^{\frac{1}{3}} \left(1 + 4\, Im(\rho_1)\right)}$

Where $s = (x + iy)$, $Im(\rho_1) = $ the imaginary part of the first nontrivial root of the Reimann zeta function ρ_1, $\log = $ the hyperbolic logarithm function, and $\Gamma(s) = $ the gamma function, which takes the integral form.

$$\Gamma(s) = \int_{0}^{\infty} \frac{x^{s-1}}{e^x - 0} dx \qquad\qquad Re(s) > 0$$

$\rho_1 = 0.5 + 14.1314251417346 \dots i$ 1st nontrivial zero of the zeta function

Special known values of the Reimann zeta function include:

$\zeta(-2k) = 0 \qquad\qquad \zeta(-2) = \zeta(-4) = \zeta(-6) = \zeta(-8) = \cdots = 0$

$\zeta(0) = -\dfrac{1}{2} \qquad\qquad \zeta'(0) = -\dfrac{1}{2}\log(2\pi)$

$\zeta(1) = \dfrac{1}{1} + \dfrac{1}{2} + \dfrac{1}{3} + \dfrac{1}{4} + \dfrac{1}{5} + \dfrac{1}{6} + \dfrac{1}{7} + \dfrac{1}{8} + \cdots = \infty$

$\zeta(2) = \dfrac{1}{1^2} + \dfrac{1}{2^2} + \dfrac{1}{3^2} + \dfrac{1}{4^2} + \dfrac{1}{5^2} + \dfrac{1}{6^2} + \dfrac{1}{7^2} + \cdots = \dfrac{\pi^2}{6}$

106

$$\zeta(3) = \frac{1}{2} \int_0^\infty \frac{x^2}{e^x - 1} dx$$

$$\sqrt{\frac{\zeta(3)}{\zeta'(-2)}} = 2\pi i$$

$$\zeta(3) = \frac{2}{3} \int_0^\infty \frac{x^2}{e^x + 1} dx$$

$$\frac{\zeta\left(\frac{3}{2}\right)}{\zeta\left(-\frac{1}{2}\right)} = -4\pi$$

$$\zeta(3) = \frac{1}{2} \int_0^\infty \frac{\log(x)\log(1-x)}{x(1-x)} dx$$

The Reimann zeta function has a built-in unitary generalized limiting dependence on the Euler-Mascheroni constant.

$$\lim_{x \to 0} \left(\frac{\zeta(1+x) + \zeta(1-x)}{2} \right) = \gamma$$

$$\lim_{n \to \infty} \left(-n + \zeta\left(\frac{n+1}{n}\right) \right) = \gamma$$

Where γ = the Euler-Mascheroni constant, and $\zeta(s)$ = the Reimann zeta function.

The Reimann zeta function generalizes to the polylogarithm function.

$$Li_s(z) = \frac{z^1}{1^s} + \frac{z^2}{2^s} + \frac{z^3}{3^s} + \frac{z^4}{4^s} + \frac{z^5}{5^s} + \frac{z^6}{6^s} + \frac{z^7}{7^s} + \frac{z^8}{8^s} + \cdots$$

$$Li_s(1) = \zeta(s)$$

$$Li_1(z) = -\log(1-z)$$

$$Li_5\left(\frac{1}{2}\right) = -\zeta(-1,-1,1,1,1)$$

Where $\zeta(s)$ = the Riemann zeta function, $\zeta(s_1, \ldots s_k)$ = the multiple zeta function, $\log(z)$ = the hyperbolic logarithm function, and $Li_s(z)$ = the polylogarithm function.

The polylogarithm function maintains golden ratio dilogarithmic connections.

$$\int_{1-\varphi}^{-\varphi} \frac{Li_2(x)}{x}\,dx = Li_3(-\varphi) - Li_3(1-\varphi)$$

$$\int_{-\frac{1}{\varphi}}^{-\varphi} \frac{Li_2(x)}{x}\,dx = Li_3(-\varphi) - Li_3\left(-\frac{1}{\varphi}\right)$$

Golden dilogarithm integrals

Using the universal binomial factorization prescription to interpret the imaginary part of the zeta function's first non-trivial zero $Im(\rho_1)$, we discover that it uniquely identifies the circular/hyperbolic factorization boundary that terminally maintains the hyperbolic vortex radius (ж_r) under simple chiral connection.

$$\text{ж}_r = Im(\rho_1)\left(\cos\left(\cosh\left(\frac{n}{2}\right)\right)\right)^{-1}\left(1 - \frac{d_2}{2}\left(\frac{n}{2}\right)Re\left(i^{i^{i^{\cdots}}}\right)\boxplus\right)$$

Where ж_r = the hyperbolic vortex radius constant, $Im(\rho_1)$ = the imaginary part of the first non-trivial zero of the zeta function, $n = 5$ the number of unique rotations partitioning the minimal arena, $Re\left(i^{i^{i^{\cdots}}}\right)$ = the real part of the infinite power tower of the imaginary unit i, $cos(x)$ = the cosine function, $cosh(x)$ = the hyperbolic cosine function, and $\boxplus$ = the terminal boundary of the minimal arena.

In other words, the imaginary part of the first non-trivial zero of the Riemann zeta function constructively defines the minimal arena's partition limit.

108

Graphing the Riemann zeta function under inverse complex argument we notice it contains split binding points, maintaining separate junctions about $(0,0)$ and $(1,0)$ (Graph 17).

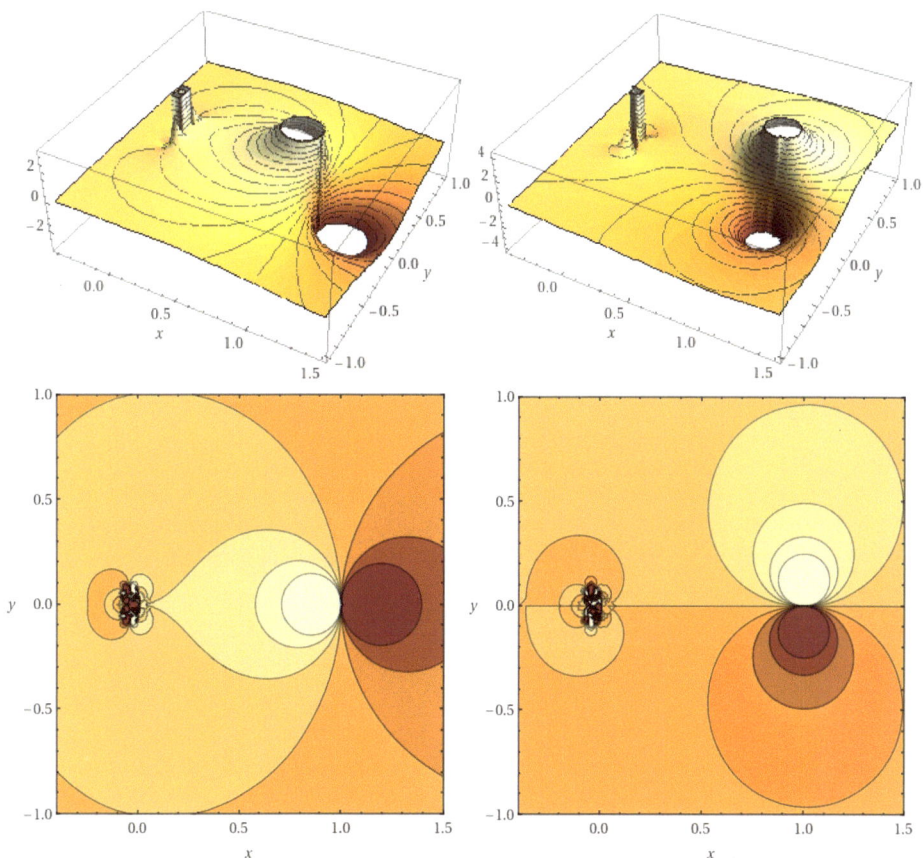

Graph 17: real (left) and imaginary (right) plots of the Riemann zeta function under inverse-complex argument $\zeta((x + iy)^{-1})$.

Zooming in on the detail around $(0,0)$ (Graph 18) gives us a reproduction of the internal action of balance 2 (Graph 6).

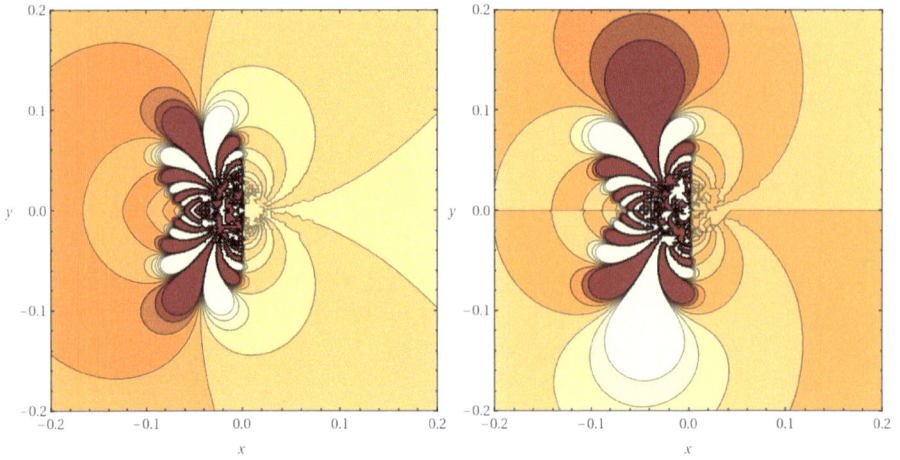

Graph 18: real (left) and imaginary (right) plots of the Riemann zeta function under inverse-complex argument $\zeta((x+iy)^{-1})$, zoomed in around $(0,0)$.

And zooming in on the detail around $(1,0)$ (Graph 19) reproduces the inverse graph of the internal action of balance 1 (Graph 3).

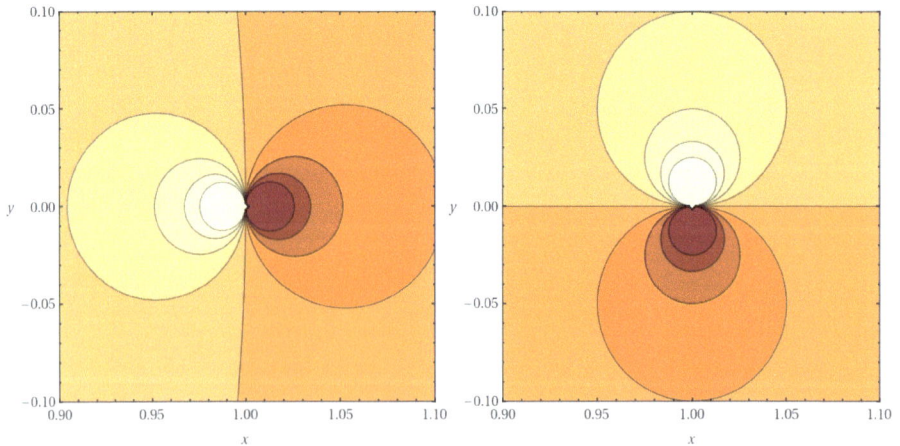

Graph 19: real (left) and imaginary (right) plots of the Riemann zeta function under inverse-complex argument $\zeta((x+iy)^{-1})$, zoomed in around $(1,0)$. Compare to Graph 3.

Having a minimal derangement/factorization theory means having a formally constructive definition for how the minimum self-divisible unitary projection (the 1) internally factors. This minimal internal factorization sets the structure in the *primes*.

---

prime limits

$$\prod_{k=1}^{\infty} \frac{p_k{}^2 + 1}{p_k{}^2 - 1} = \frac{n}{2} \qquad \text{infinite product of primes}$$

$$\lim_{n\to\infty} \frac{1}{\log p_n} \prod_{k=1}^{n} \frac{1}{1 - \dfrac{1}{p_k}} = e^{\gamma} \qquad \text{inverse log product limit}$$

$$\lim_{n\to\infty} \log p_n \prod_{k=1}^{n} \frac{1}{1 + \dfrac{1}{p_k}} = \frac{\zeta(2)}{e^{\gamma}} = \frac{\pi^2}{6\,e^{\gamma}} \qquad \text{log product limit}$$

Where p is a prime number, $n = 5$, π = Archimedes' constant, γ = the Euler-Mascheroni constant, $\log(x)$ = the hyperbolic logarithm function, and $\zeta(s)$ = the Reimann zeta function.

For a richer understanding of the zeta function we can use the universal binomial factorization prescription to ask many new questions. For example,

Question 1: What is the zeta function value of the time boundary's rotation?

$$\zeta(\phi_0) = 3\,Re(\omega_1)^4 \left(1 + 3\,\Gamma(n)\left(\omega_2\,\Gamma(x_{min})\right)^4 \boxplus\right)$$

Where ϕ_0 = the rotation of the time boundary, ω_1 and ω_2 = the omega_1 and omega_2 constants, $\zeta(x)$ = the zeta function, $\Gamma(x)$ = the gamma function, $\Gamma(x_{min})$ = the minimum value of the gamma function for positive argument, and $\boxplus$ = the terminal boundary of the minimal arena.

Question 2: What is the zeta function value of the space boundary's rotation?

$$\zeta(\phi_1) = 3\, m_R \left(1 + \left(\frac{\Gamma(n)}{!n\, W(1)^4}\right)^3 \text{Ж}_r{}^2\ \boxplus\right)$$

Where ϕ_1 = the external rotation of the space boundary, Ж_r = the hyperbolic vortex radius constant, $W(1)$ = the omega constant—the unitary value of Lambert's W function, defined as the unique real number that satisfies the equation $\Omega e^{\Omega} = 1$, m_R = Rényi's parking constant, $\Gamma(x)$ = the gamma function, $\zeta(x)$ = the zeta function, and $\boxplus$ = the terminal boundary of the minimal arena.

$$m_R = e^{-2\gamma} \int_0^{\infty} \frac{e^{-2\Gamma(0,x)}}{x^2} = e^{-2\gamma} \int_0^{\infty} \frac{e^{2Ei(-x)}}{x^2}$$

Question 3: What is the zeta function value of the Planck length's inverse square rotation?

$$\zeta(\phi_1{}^{-2}) = -\left(\frac{2}{3}\, e^{\gamma}\right)^{\frac{1}{2}} \left(1 + 2n \left(3V_{fe}\right)^2 \boxplus\right)$$

Where ϕ_1 = the external rotation of the space boundary, e = Euler's number, γ = the Euler-Mascheroni constant, V_{fe} = the volume of the hyperbolic figure eight knot complement, $\zeta(x)$ = the zeta function, and $\boxplus$ = the terminal boundary of the minimal arena.

Question 4: What is the zeta value of inverse hyperbolic vortex radius division?

$$\zeta\left(\frac{1}{\text{Ж}_r}\right) = -4\pi\, \sinh^{-1}\left(\frac{1}{4}\right)^2 \left(1 + \left(2\, \text{Ж}_2\, G_g\right)^2 \boxplus\right)$$

Where π = Archimedes' constant, Ж_2 = the 2nd hyperbolic vortex partition constant, G_g = the tether length at which a goat tied to the boundary of a unit circular field can graze exactly half the field, $\sinh(x)$ = the hyperbolic sine function, and $\boxplus$ = the terminal boundary of the minimal arena.

Question 5: What is the zeta value of square inverse hyperbolic vortex radius division?

$$\zeta\left(\frac{1}{\text{Ж}_r{}^2}\right) = -\frac{d_1}{C_{Murata}{}^4}\left(1 + \frac{\pi}{8}\cosh^2(4)\ \boxplus\right)$$

Where π = Archimedes' constant, Ж_r = the hyperbolic vortex radius, $\cosh(x)$ = the hyperbolic cosine function, d_1 = the 35 derangements of the space boundary, and C_{Murata} = the Murata constant defines the following product over the primes.

$$C_{Murata} = \prod_p \left(1 + \frac{1}{(p-1)^2}\right)$$

Where p = a prime number.

Chapter 13: internal and external logic

There are exactly 4 kinds of numbers that can be added, subtracted, multiplied and divided. In other words, there are exactly 4 division algebras: real numbers, complex numbers, quaternions, and octonions. The real numbers are 1-dimesnional. Pairs of real numbers make the complex numbers, which are 2-dimensional. Likewise, pairs of complex numbers make the quaternions (4-D), and pairs of quaternions make the octonions (8-D). And that's the end of the line. That is, the octonions represent the last of the division algebras. No higher order pairings will allow all those symmetric operations.

arithmetic arenas	
real ($\mathbb{R}$)	1D
complex ($\mathbb{C}$)	2D
quaternion ($\mathbb{H}$)	4D
octonion ($\mathbb{O}$)	8D

These division algebras reflect the closed division balances of the minimal arena's different constructive domains. The internal domain closes under 8D octonion connection, whereas the external domain rearranges those same parts to close under 4D quaternion connection, which defines a union of complex planes intersecting on the real line.

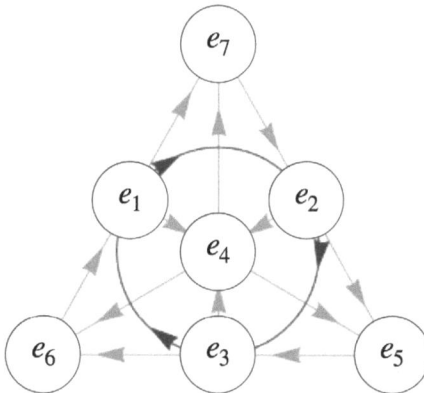

Cayley logic table for the octonions portraying the 7 basis vectors e_k and their algebraic connections.

Since the octonion logic symmetrically connects 8 things (7 basis vectors plus the external unitary projection 1) under closed 3-member operations, it topologically decomposes into 8 simply connected triangles, or 2 tetrahedra. Recall that the hyperbolic figure eight knot decomposes into a union of 2 regular ideal hyperbolic tetrahedra, and a tetrahedron decomposes into 4 simply connected triangles.

To visualize the 8 symmetric triple groupings of the minimal arena's internal octonion connection we graph the product of the inner action of balance 0 and balance 1.

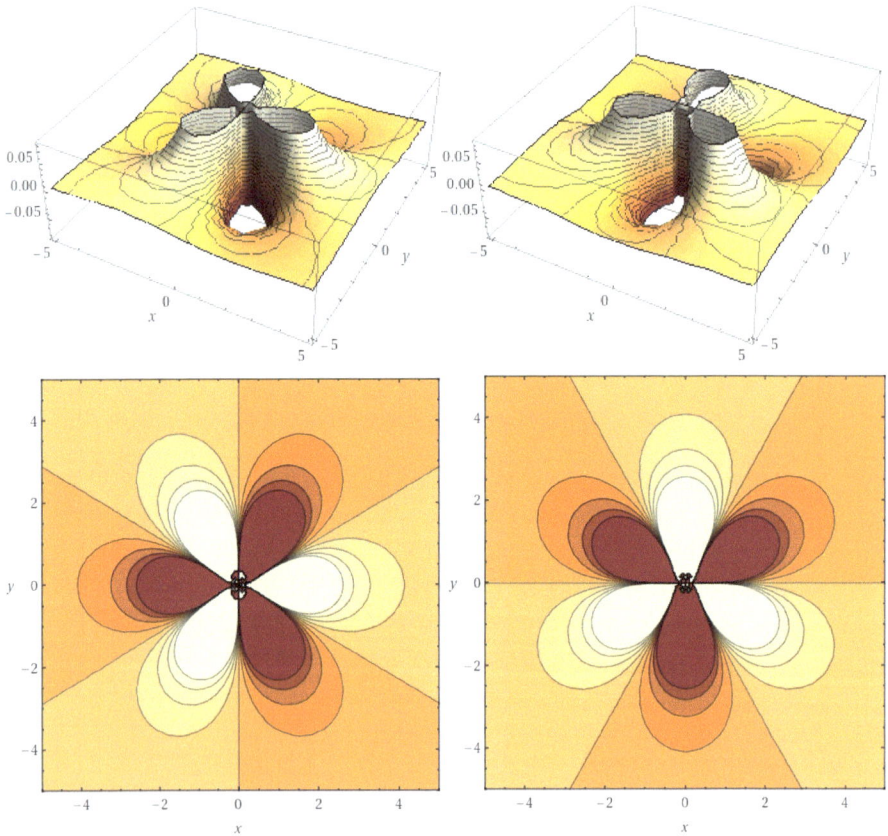

Graph 20: real (left) and imaginary (right) plots of the product of the internal actions of balance 0 and 1, each under inverse-complex argument. Compare to the Cayley logic table for the octonions.

Both the *real* and *complex* graphs identify 2 inner groups of 3 and 2 outer groups of 3 that connect under 5-part asymmetric balance; defining the 8 3-member simply connected closed operations of the octonion logic.

116

The minimal arena is internally bound into the hyperbolic figure eight knot under octonion connective logic, and externally bound into the n-hypersphere of maximal volume under quaternion connective logic. This quaternion connection is supersymmetrically coded by the fact that every possible labeling (a, b, c, d) of $\{\, Re(ж_3), Im(ж_3), Re(ж_4), Im(ж_4)\,\}$ the real and imaginary parts of the 3rd and 4th hyperbolic vortex partition constants satisfies the quaternion algebraic identity.

$$(ac - bd)^2 + (ad + bc)^2 = (a^2 + b^2)(c^2 + d^2)$$

Paired labelings (those that group real and imaginary parts together) constructively define the hyperbolic vortex radius to the 4th power.

$$(ac - bd)^2 + (ad + bc)^2 = (a^2 + b^2)(c^2 + d^2) = ж_r{}^4$$

Where $ж_r$ = the hyperbolic vortex radius, and (a, b), (c, d) pair real and imaginary parts of $ж_3$ and $ж_4$.

Another baked in consequence of the minimal arena's external shape is *Calculus*. In addition to the fact that the generalized power series of the external domain's hyperbolic connection (e^x) defines the elementary derivative sequence,

$$e^x = \frac{x^0}{0!} + \frac{x^1}{1!} + \frac{x^2}{2!} + \frac{x^3}{3!} + \frac{x^4}{4!} + \frac{x^5}{5!} + \cdots$$

The hyperbolic power series and the derivative. Every term in the generalized exponential series defines the derivative of the following term.

the *derivative* geometrically defines how the volume of a n-dimensional hypercube increases, or decreases, as the side length of that hypercube is changed.[7]

---

[7] The exponential function is also the unique nontrivial function that is its own derivative (up to multiplication by a constant) and its own

On the other hand, "integrating this picture—stacking the faces—geometrizes the fundamental theorem of calculus, yielding a decomposition of the n-cube into n pyramids, which is a geometric proof of Cavalieri's quadrature formula."[8]

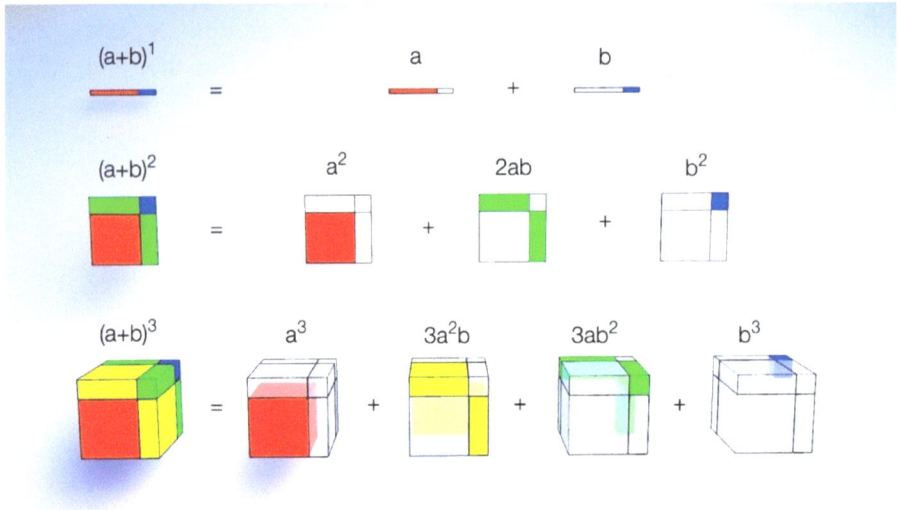

A geometric proof of the derivative and integral, a visualization of the binomial expansion up to the 3rd power.

What the minimal arena's external projection looks like depends on the scale we focus on. On its grandest scale (1.41678698590795 ... $\times$ 10^{32} meters) the n-hypersphere of maximal volume defines the simplest possible geometric form in $\mathbb{S}^3$—*the great sphere* minus its boundary.

On its smallest scales (between 1.87554596713962 ... $\times 10^{-18}$ meters and 2.17642683817579 ... $\times 10^{-8}$ meters) it defines the second simplest geometric form, the minimal hypersurface known as the Clifford torus.[9]

---

antiderivative. In the graph $y = e^x$ the slope of the curve at any point x is e^x, and the area under the curve from $-\infty$ to x is also e^x.

$$\frac{d}{dx}e^x = e^x \qquad \int e^x \, dx = e^x + C$$

[8] Cavalieri's quadrature formula Wikipedia page. Note: the area, or "quadrature", of a hyperbola defines a logarithm.
[9] The Clifford torus is a Möbius torus. Going around the boundary once puts you on its other side of its surface, requiring 2 loops to return to the

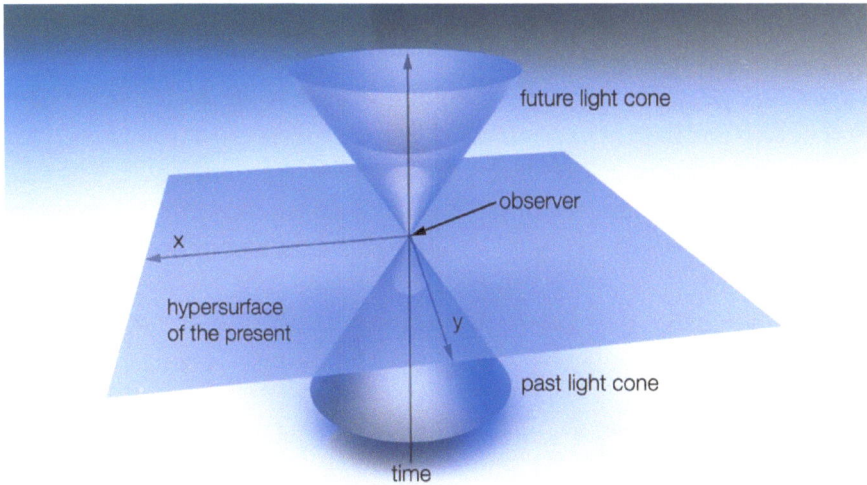

The projective tangent cone of the Clifford torus.

The tangent cone of the Clifford torus defines the projective *null cone* of hyperbolic geometry. The entire surface of this hypercone defines the division singularity of the external domain. And the balance maintained by this constructive arrangement (between the future hypercone and the past hypercone) defines the *hyperplane* of the present—the most trivial minimal hypersurface.

As a helicoidal product surface, the Clifford torus defines the simplest and most symmetric flat embedding of the cartesian product of two circles. That is, it defines the simplest connection between 2 circles, each of which possess their own independent embedding space $\mathbb{R}^2_a$ and $\mathbb{R}^2_b$, resulting in a product space that is $\mathbb{R}^4$ (the 4-dimensional hyperbolic spacetime of general relativity).[10]

---

starting condition (see Heegaard splitting—embedded minimal surfaces in compact manifolds of positive sectional curvature). The Clifford torus is also an example of a square torus, because it is isometric to a square with opposite sides identified (making it a manifold). The square torus can also be embedded into three-dimensional space by the Nash embedding theorem. One possible embedding modifies the standard torus by a fractal set of ripples running in two perpendicular directions along the surface.

[10] Since $\mathbb{R}^4$ is identified with $\mathbb{C}^2$, this is equivalent to saying that the 3-sphere $\mathbb{S}^3$ lives in $\mathbb{C}^2$.

But that's not the end of the story. Since this helicoidal surface connects 2 circles of significantly different size, the chiral connection of its rotation operator is highly asymmetric, causing the cartesian product of these two circles to twist into an $\mathbb{R}^3$ torus.

Under asymmetric chiral connection this is unavoidable. The first circle in that connection consumes x and y leaving only one independent axis z available to the second circle. This quickly collapses the $\mathbb{R}^4$ domain into a $\mathbb{R}^3$ projection. "A torus embedded in $\mathbb{R}^3$ is an asymmetric reduced-dimension projection of the maximally symmetric Clifford torus embedded in $\mathbb{R}^4$."[11]

The generalized equation for the tangent cone of the Clifford torus is the four-dimensional relation:

$$x_1 x_2 = x_3 x_4$$

Where x_k = the kth factor (dimension) of its connection.

The null cone of the n-hypersphere of maximal volume is set by the 4 rotations of the minimal arena's terminal division boundaries.

$$\phi_1 \phi_3 = \phi_2 \phi_2$$

Where ϕ_1, ϕ_2, and ϕ_3 = the rotations of the Planck length, charge, and mass boundaries, and $\boxplus = \left(\dfrac{l_p \, m_p}{q_p{}^2} \right)$ the terminal boundary of the minimal arena.

This null cone defines the Clifford torus with maximum square symmetry. That is, for any possible set of different sized circles put into Clifford torus arrangement, the product of the left pair of operators equals the product of its right pair of operators. But for one supersymmetric Clifford torus there exists an additional square symmetry, where one of its paired sets of operators also share an equality between them. The null cone of this maximally square symmetric Clifford torus defines the projective division boundary of the n-hypersphere of maximal volume.

---

[11] https://en.wikipedia.org/wiki/Clifford_torus

The other minimal hypersurfaces embedded in this geometry (those that play a role in the helicoidal collapse from $\mathbb{R}^4$ to $\mathbb{R}^3$) are found by generalizing the Clifford torus via the method of separation of variables.[12]

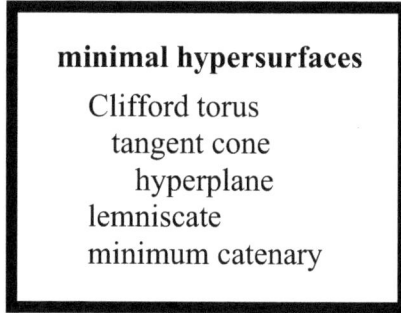

> **minimal hypersurfaces**
>
> Clifford torus
> tangent cone
> hyperplane
> lemniscate
> minimum catenary

"These are all the minimal hypersurfaces that can be obtained by the method of separation of variables."

Jaigyoung Choe

---

[12] See Jaigyoung Choe. Some minimal submanifolds, generalizing the Clifford torus.

Chapter 14: hyperbolic

In this book we have found that the minimum volume complement (the hyperbolic figure eight knot) and its external inverse (the n-hypersphere of maximal volume) jointly form the minimal arena, maintaining 5 perpetual actions (time, space, charge, mass and temperature) with Planck constant boundaries. Under universal binomial factorization the external *charge* and *mass* boundaries of this minimal arena partition into the exact charge and mass values that define the fundamental particles of matter, and the unique 44 derangements of this 5-dimensional arena define the constants of Nature.

We also found that this minimal geometry arranges the Planck boundaries into a dually ideal hyperbolic union, defined by Euler's number e, and the gamma function. And that Calculus is scripted by how the external boundaries of the minimal arena partition (defining differentiation), and seamlessly integrate back together (defining integration).

In short, the simplest manifold (the simplest possible self-closed geometric balance) logically constructs the minimal stage of persistence, defining the base combinatorial structure of physical reality and the base logic of persistent self-division.

This insight into the inner workings of Nature has the power to elevate us into a world in which money is irrelevant. And this, I believe, warrants our utmost attention.

Since it is the case that every unique persistent form in Nature (a strawberry, a brick of gold, an ice cube, …) is a balanced composition of charge boundaries, available in only one magnitude (e = the magnitude of the electron charge) and mass boundaries, available in 17 magnitudes (m_e, m_+, m_N, m_μ, m_τ, m_Z, m_W, m_H, m_t, m_c, m_b, m_s, m_d, m_u, m_{ν_τ}, m_{ν_μ} and m_{ν_e} = the masses of the: electron, proton, neutron, muon, tau, Z boson, W boson, Higgs boson, truth (top) quark, charm quark, beauty (bottom) quark, strange quark, down quark, up quark, tau neutrino, muon neutrino, and the electron neutrino), there exists a simple logic encoding the minimal transformations between every stable set of combinations.

Compared to this fact, nothing in humanity's current political landscape holds any value. Nothing we are fighting over matters in the slightest.

Every form of food, every object ever held in a hand, any resource has ever been interacted with, or ever could be, is simply a composition of charge and mass boundaries. Knowing the shape of those boundaries and how they are logically constrained makes it possible to use them as tools. The ability to leverage Nature's inherent boundaries to transform material things to other material things (up to the same magnitude) will change the course of the human day more than any other trans framing has.

When the simplest way to acquire every resource is to convert the local matter you have into the form of matter you want, business becomes a thing of the past. Every individual becomes their own fountain of resources. No child is stopped from exploring, tinkering, learning, and continuing their quest to grow up throughout life due to a lack of resources. No large overarching systems are needed to oversee the direction and distribution of resources. No middle men are needed.

With replicator technology there is no point in fighting over resources. When everyone has complete access to resources on their own, when we are limited only by our ability to conceive and communicate our wants, the concepts of power, war, greed, envy, simply dilute away. Class distinctions become meaningless. Large overarching control structures become entirely unneeded and disappear, not by being violently overthrown, or replaced, but rather, by becoming entirely irrelevant.

The focus on money guarantees society rot. When every mind is encouraged first and foremost to sharpen their talent at making money, the consequences are inevitable. Some of those bright minds succeed. They figure out more ways of reliably getting the attention of larger and larger groups of people, and they get rich along the way. And the rest of the industry copies them.

Everyone quickly learns that making money is first and foremost a game of getting the attention of the largest number of people as possible, and only second a game of manipulating them to spend their money. As a consequence, the focus on money guarantees that we live in a world where the most talented minds take on the task of getting ahold of our attention. And it also guarantees that none of them are actually trying to warrant our attention. "Its just business," they say, a message that makes people shrink in fear always results in more money than a message that lifts the receiver.

It doesn't matter if every businessperson starts out telling him/herself that they are not going to resort to the ugly practice of stealing people's attention undeservingly, but will instead devote
124

themselves to making money in a way that deserves the attention of others with a product that genuinely improves lives. The loud, startling, violent, threatening, worry causing, burdening… approach always gets the first response, always steal the attention. Therefore, every businessperson must make a choice between not making money, or becoming another person vying for the undeserved attention of the masses. Nothing in the history of human existence has wholly consumed and utterly wasted the experience of living more than this. Nothing has shrunk the reach of human consciousness, or stunted the joyous breadth of human life more than the focus on money.

A world with replicators returns us to a world where sentience directs its own attention.

Acknowledgements

I thank George Cassiday for the passion of his ETI class.

I thank Albert Einstein for his courageous aim, Leonhard Euler for his ongoing transformative insights, and Gene Shoemaker (the man on the moon) for his bold example.

I thank my Num for pinky holds, Shangri-La, and for really meaning it.

I thank my $\sqrt{-1}$ Cloud for her delightful imagination.

I thank Elaine, Phil, and Matt Emmi for the way they gave me a key.

I thank Hans de Vries for his observation, and Philipp Preetz for pointing me to it.

I thank Archimedes, Leonard Euler, Albert Einstein, Michael Faraday, Benoit Mandelbrot, Louis de Broglie, David Bohm, Richard Feynman, Stephen Wolfram, Norman Wildberger (Insights into Mathematics), Burkard Polster (Mathologer), Shelly Goldstein, Detlef Dürr, Nino Zanghí, Robert Brady, Ross Anderson, Erwin Madelung, Grant Sanderson (3Blue1Brown), Carl Friedrich Gauss, Felix Klein, Henri Poincaré, Franck Laloë, William Thomson, Grigori Volovik, Anaximander, Yoshio Koide, Derek Muller (Veritasium), Nikolai Lobachevsky, Eugenio Beltrami, János Bolyai, Tim Maudlin, Craig Callendar, Christian Wüthrich, Destin Sandlin (Smarter Every Day), Dianna Cowern (Physics Girl), Cohl Furey, Garrett Lisi (PSI), TED, The Royal Institution, Brady Haran (Numberphile), David Richeson, Michael Atiyah, and Jaigyoung Choe—for their priceless gifts of clarity.

I thank Henry Segerman for his hyperbolic figure eight knot STL artwork.

I thank WolframAlpha.com and Wikipedia.com for their infinite usefulness in the process of inquiry.

And I thank Steve Desofi, Kevin Sirois, David Heggli, Avi Ruben, and Jerry Gardner for supporting this research.

Appendix a: graphs

To develop an intuitive sense of the rich fractal nature of these functions, I recommend plotting the graphs in this book over a wide range of scales. They can be plotted at WolframAlpha.com by entering the following.

Graph 1: plot pi(sinh((1/2*1/(x+iy))^2))^2, {x,-1.5,1.5}
Graph 2: plot pi(sinh((1/2*(x+iy))^2))^2, {x,-6,6}
Graph 3: plot (sinh(sinh(1/7*(x+iy))))^(-1)
Graph 4: plot (sinh(sinh(1/7*1/(x+iy))))^(-1), {x,-.042,.042}
Graph 5: plot 5/((7pi)^(1/2)*(3^(1/3)))*(2^10*e^pi)^(-1/8)
       gamma(1/4(x+iy))^2/(gamma(1/2*(x+iy)))
Graph 6: plot 5/((7pi)^(1/2)*(3^(1/3)))*(2^10*e^pi)^(-1/8)
       *gamma(1/4*1/(x+iy))^2/(gamma(1/2*1/(x+iy))), {x,-.1,.1}
Graph 7: plot 2pi*5*(cos(7/5*1/(x+iy)))^2, {x,-8,8}
Graph 8: plot 2pi*5*(cos(7/5*(x+iy)))^2, {x,-5,5}
Graph 9: plot (2*cosh(log(7*(x+iy)))*(cosh(5/2*(x+iy)))^2
       (cos(7/5(x+iy)))^2), {x,-4,4}
Graph 10: plot (2*cosh(log(7*1/(x+iy)))*(cosh(5/2*1/(x+iy)))^2
       *(cos(7/5*1/(x+iy)))^2), {x,-3,3}
Graph 11: plot gamma(x+iy), {x,-8,8}
Graph 12: plot gammaRegularized(1,1/(x+iy))
Graph 13: plot gammaRegularized(1,1/(x+iy)), {x,-.25,.25}
Graph 14: plot gammaRegularized(1,(x+iy))
Graph 15: plot gammaRegularized(1,(x+iy),1/(x+iy))
Graph 16: plot zeta(x+iy), {x,-15,15}
Graph 17: plot zeta(1/(x+iy)), {x,-.4,1.5}
Graph 18: plot zeta(1/(x+iy)), {x,-.2,.2} {y,-.2,.2}
Graph 19: plot zeta(1/(x+iy)), {x,.9,1.1} {y,-.1,.1}
Graph 20: plot pi*(sinh(1/2*1/(x+iy))^2)^2
       *[(sinh(sinh(1/7*1/(x+iy))))^(-1)], {x,-5,5}

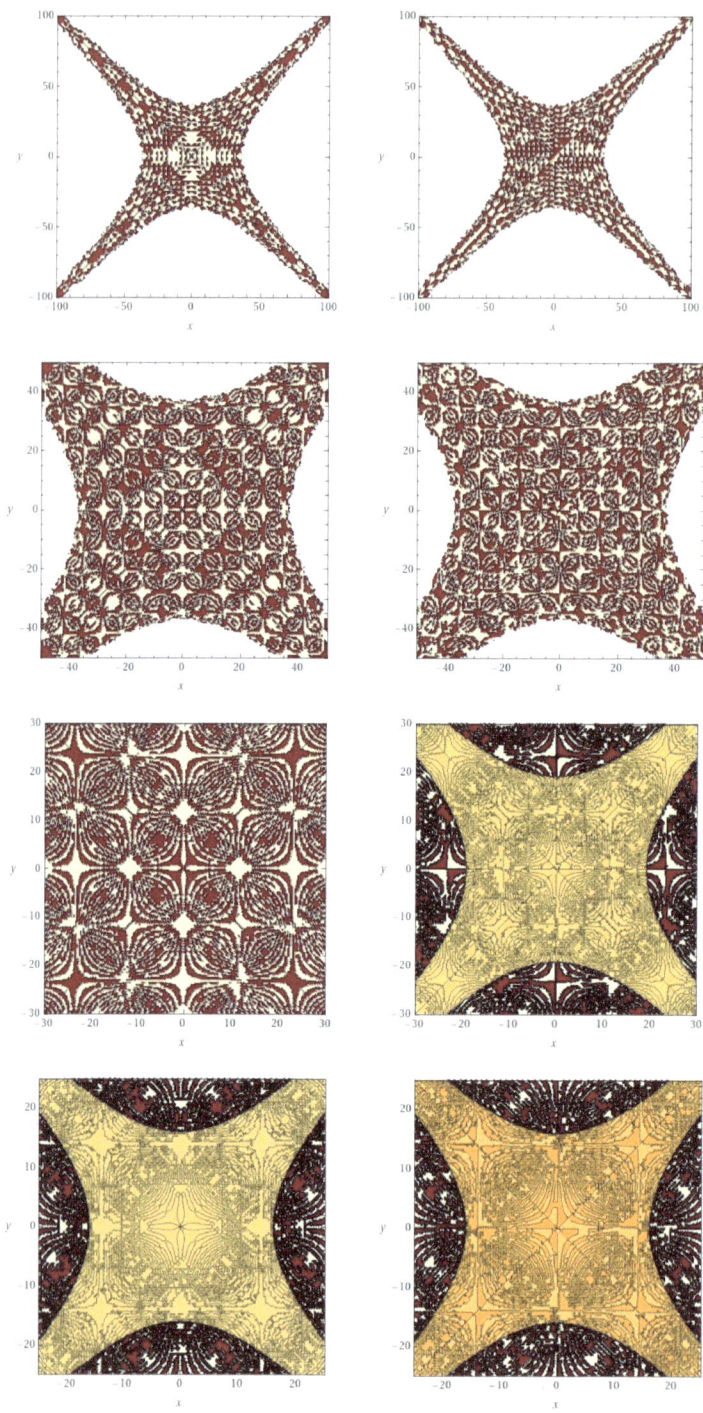

Real (left) and imaginary (right) plots of the inner action of balance 0 over a range of scales, under complex argument.

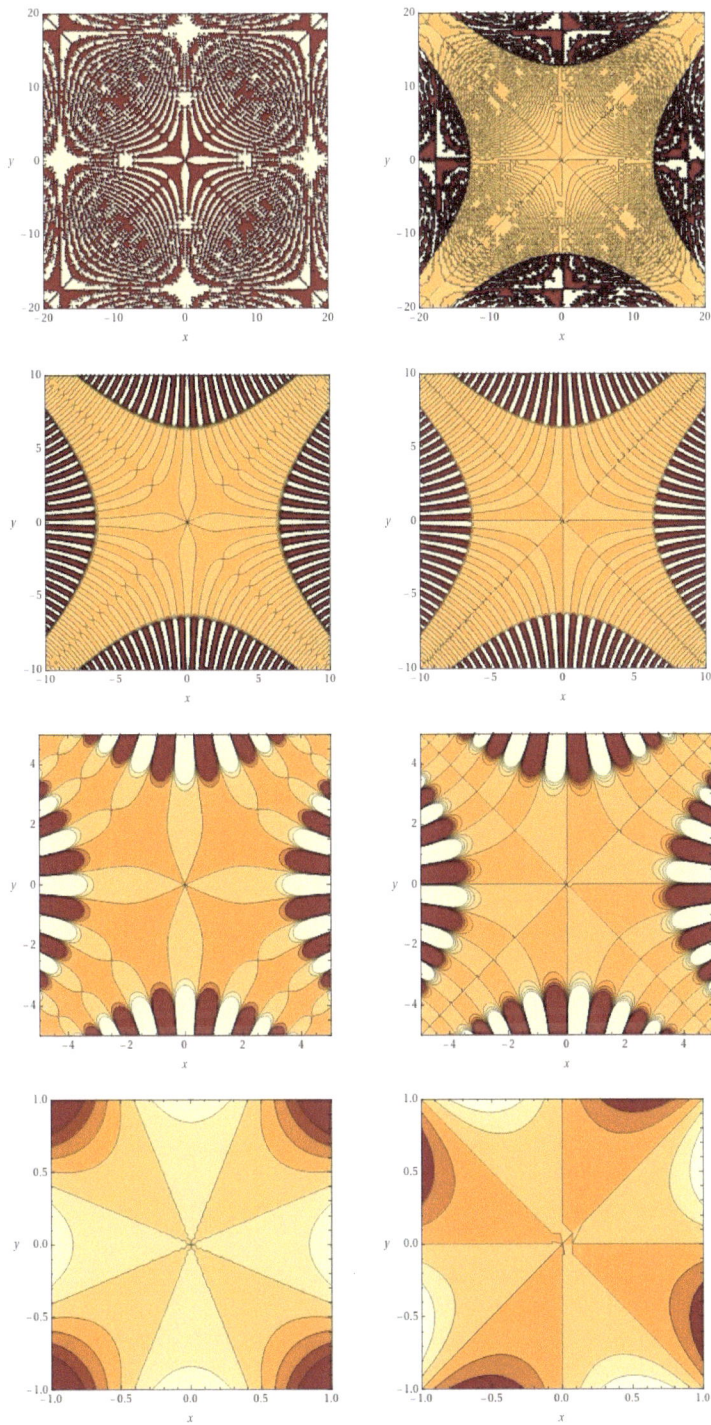

Real (left) and imaginary (right) plots of the inner action of balance 0 over a range of scales, under complex argument (continued).

131

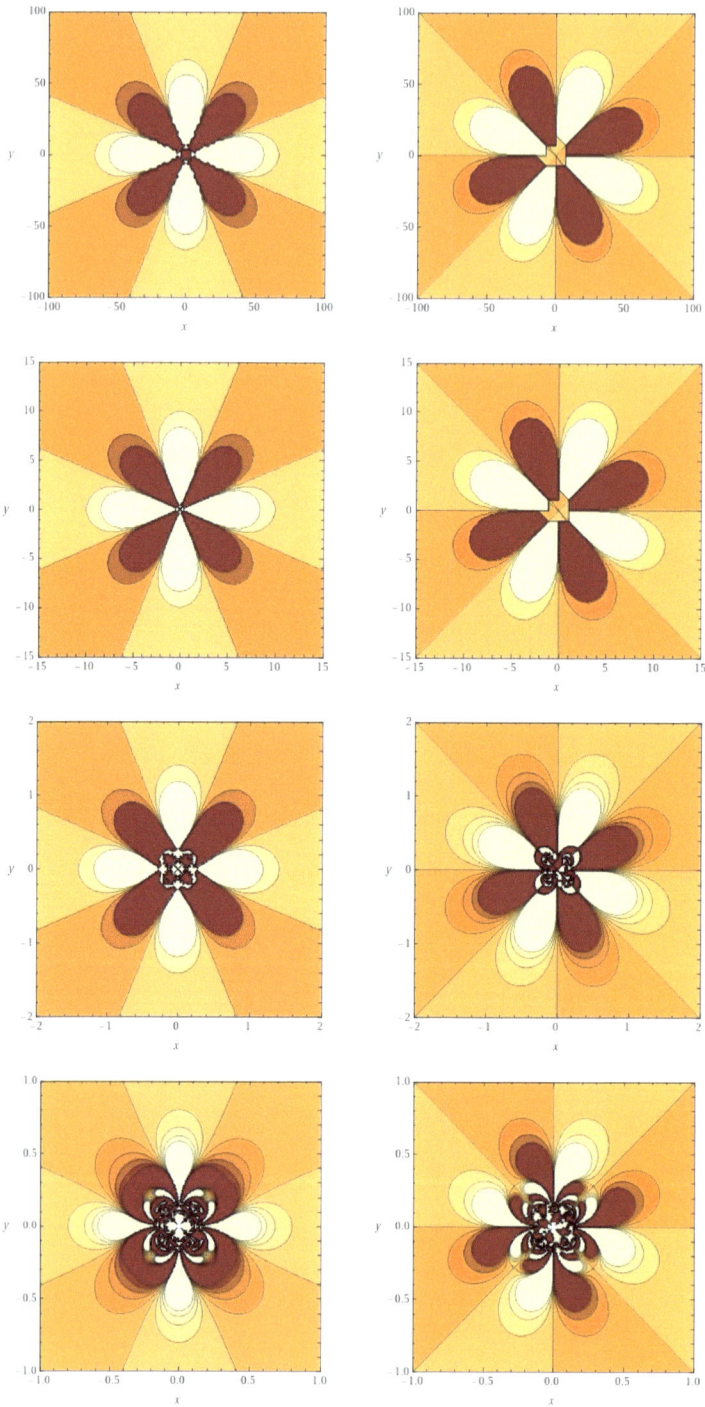

Real (left) and imaginary (right) plots of the inner action of balance 0 over a range of scales, under inverse-complex argument.

Real (left) and imaginary (right) plots of the inner action of balance 0 over a range of scales, under inverse-complex argument (continued).

133

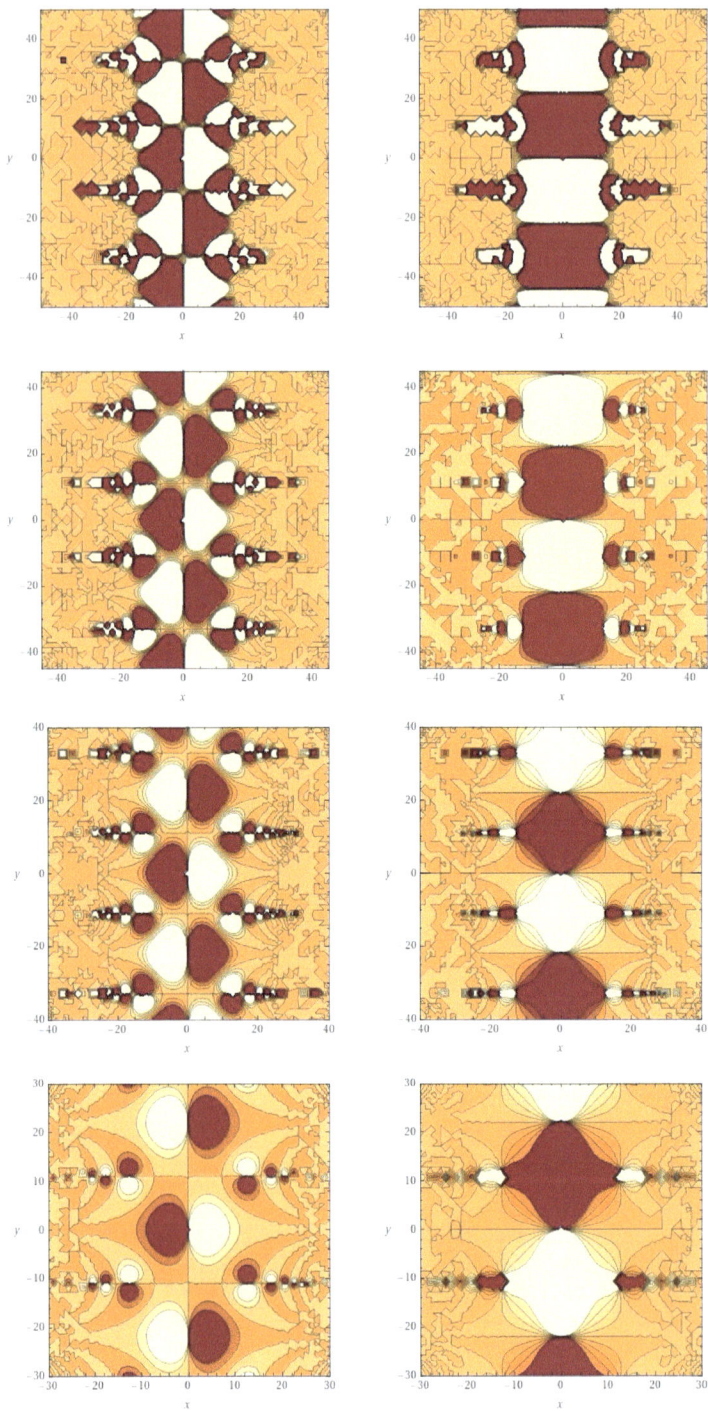

Real (left) and imaginary (right) plots of the inner action of balance 1 over a range of scales, under complex argument.

134

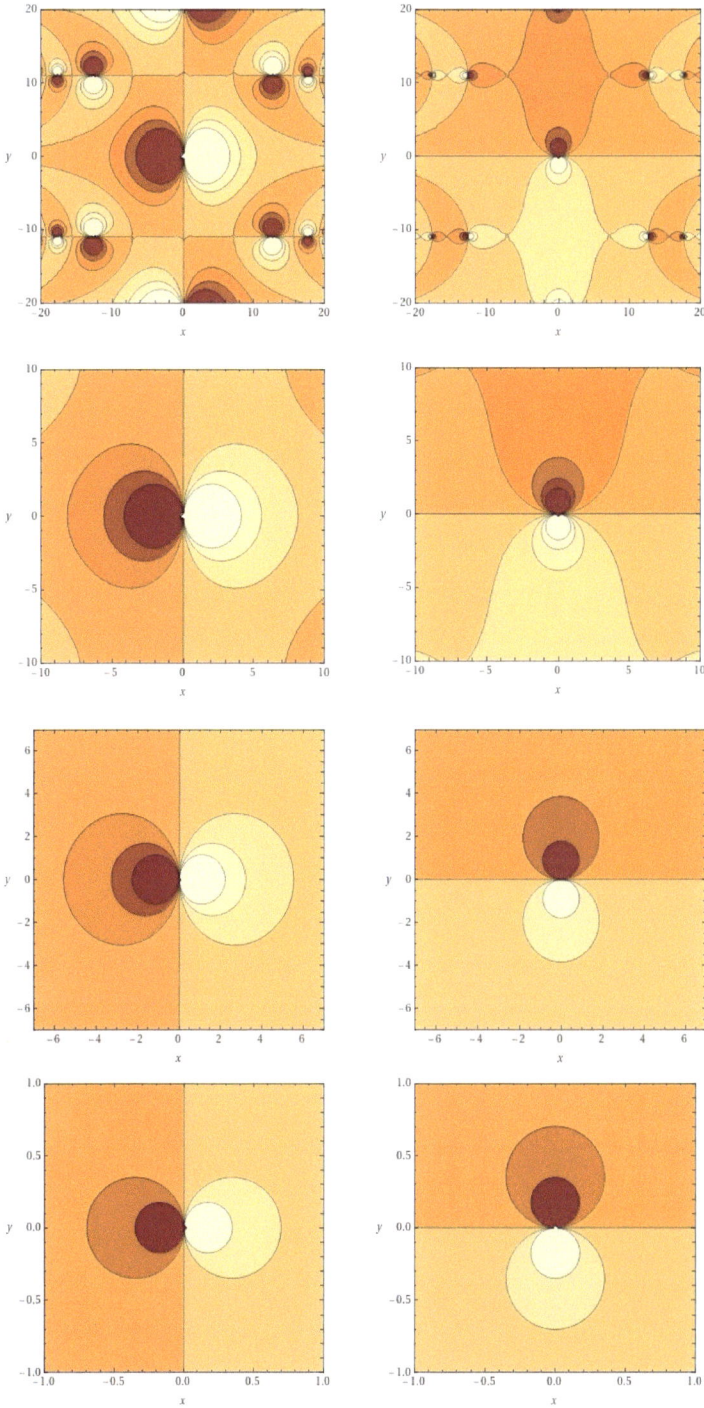

Real (left) and imaginary (right) plots of the inner action of balance 1
over a range of scales, under complex argument (continued).

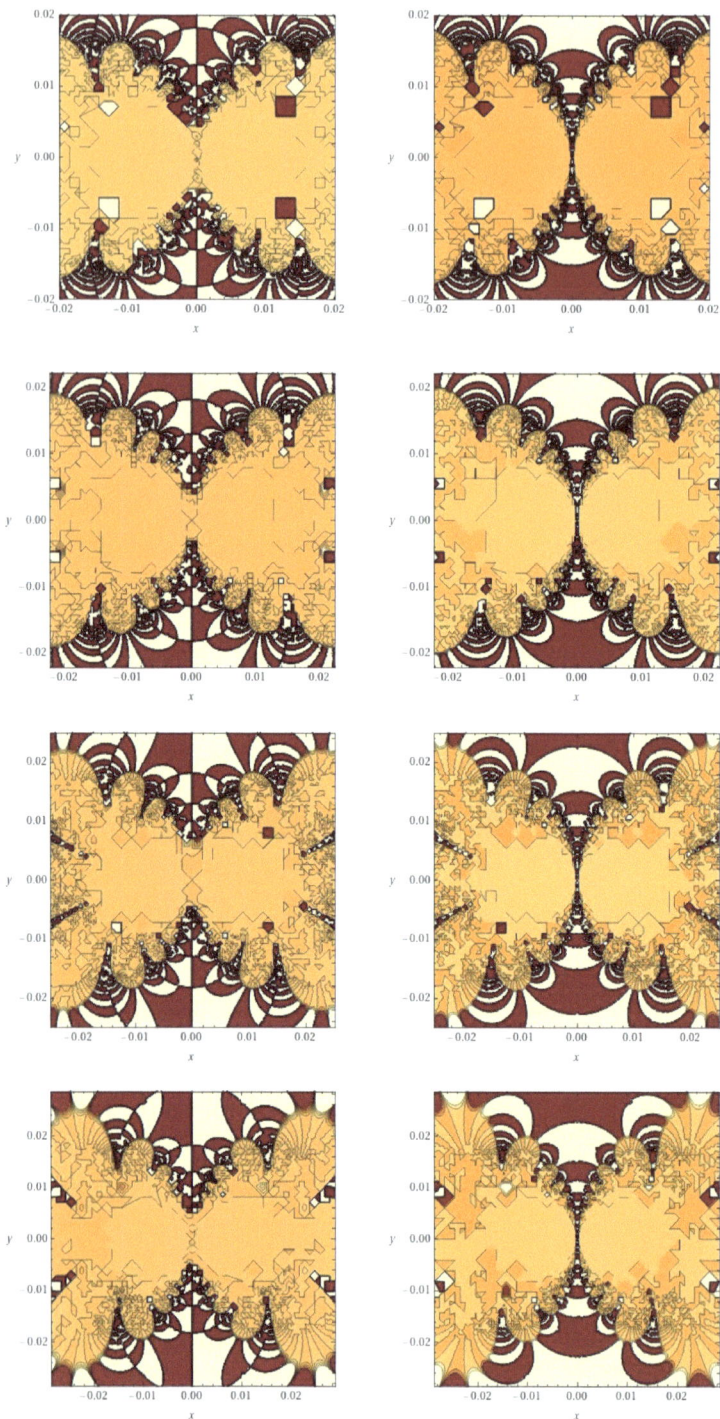

Real (left) and imaginary (right) plots of the inner action of balance 1 over a range of scales, under inverse-complex argument.

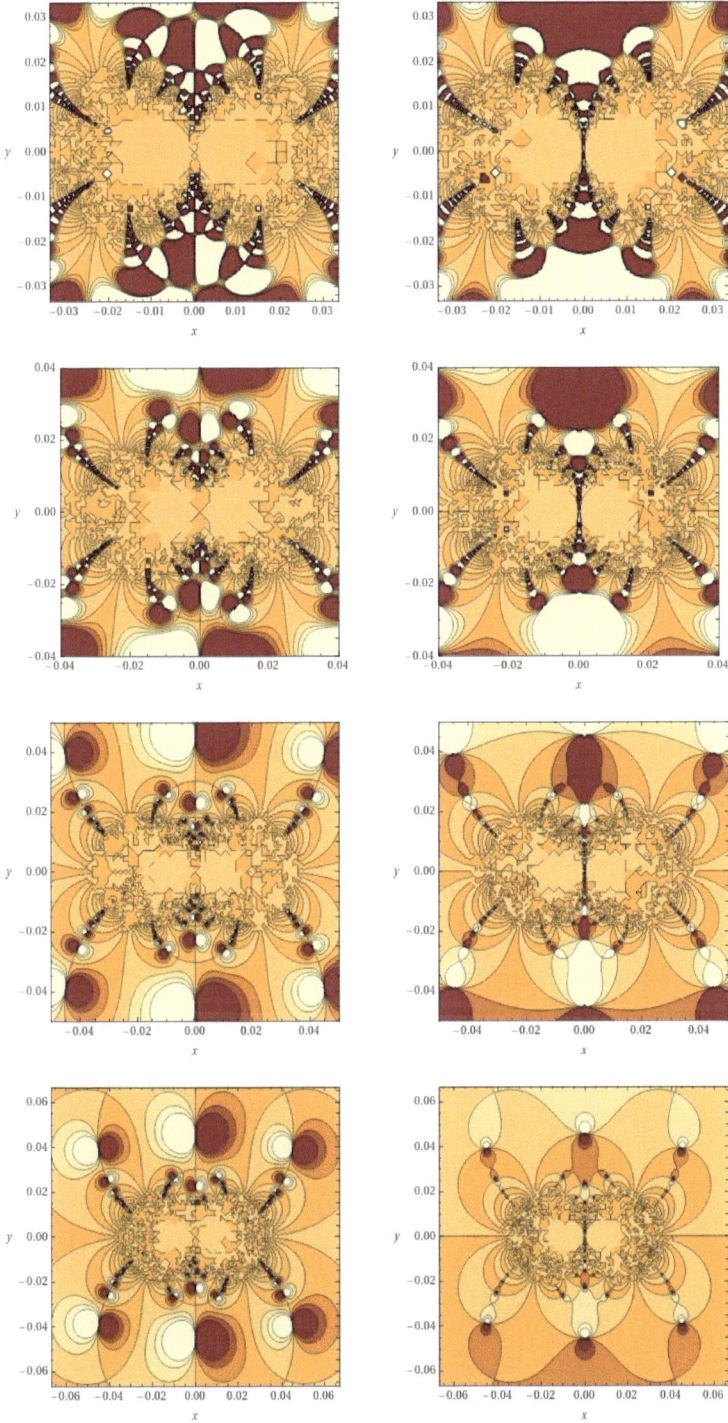

Real (left) and imaginary (right) plots of the inner action of balance 1 over a range of scales, under inverse-complex argument (continued).

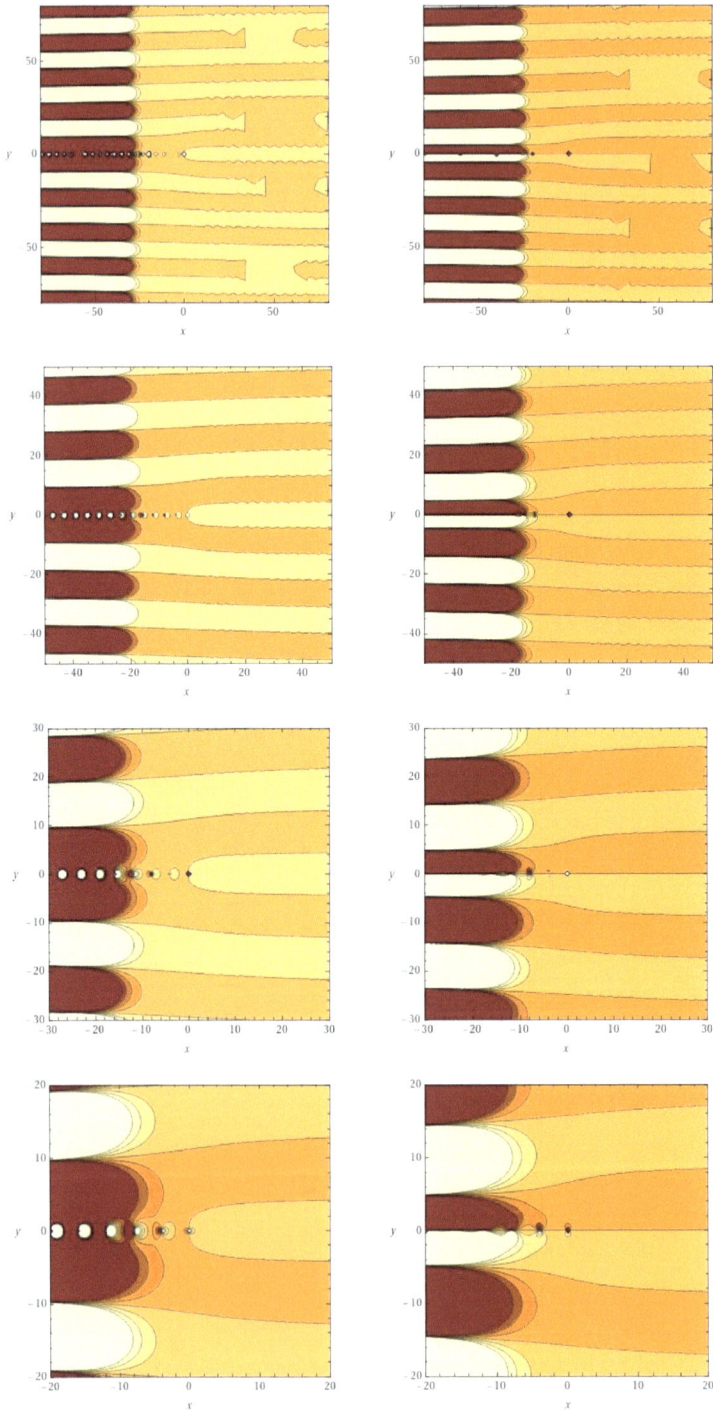

Real (left) and imaginary (right) plots of the inner action of balance 2 over a range of scales, under complex argument.

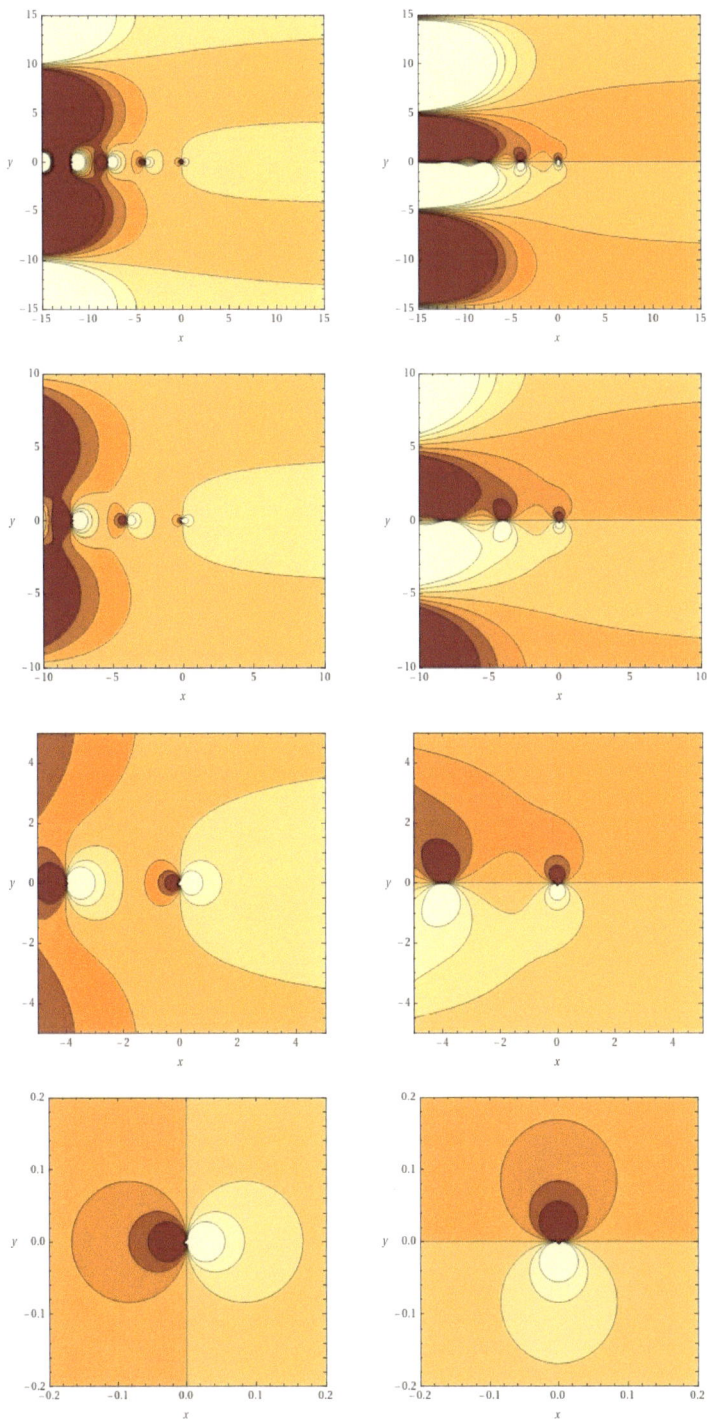

Real (left) and imaginary (right) plots of the inner action of balance 2 over a range of scales, under complex argument (continued).

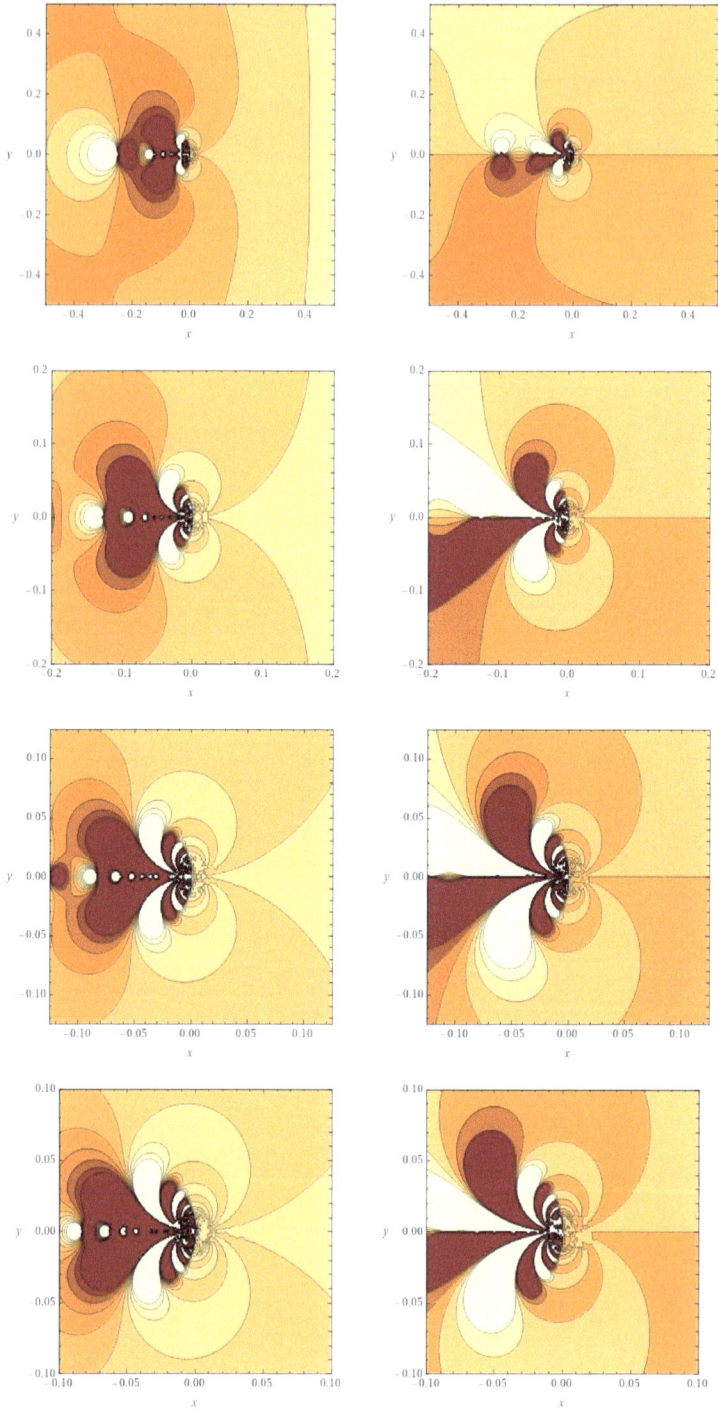

Real (left) and imaginary (right) plots of the inner action of balance 2 over a range of scales, under inverse-complex argument.

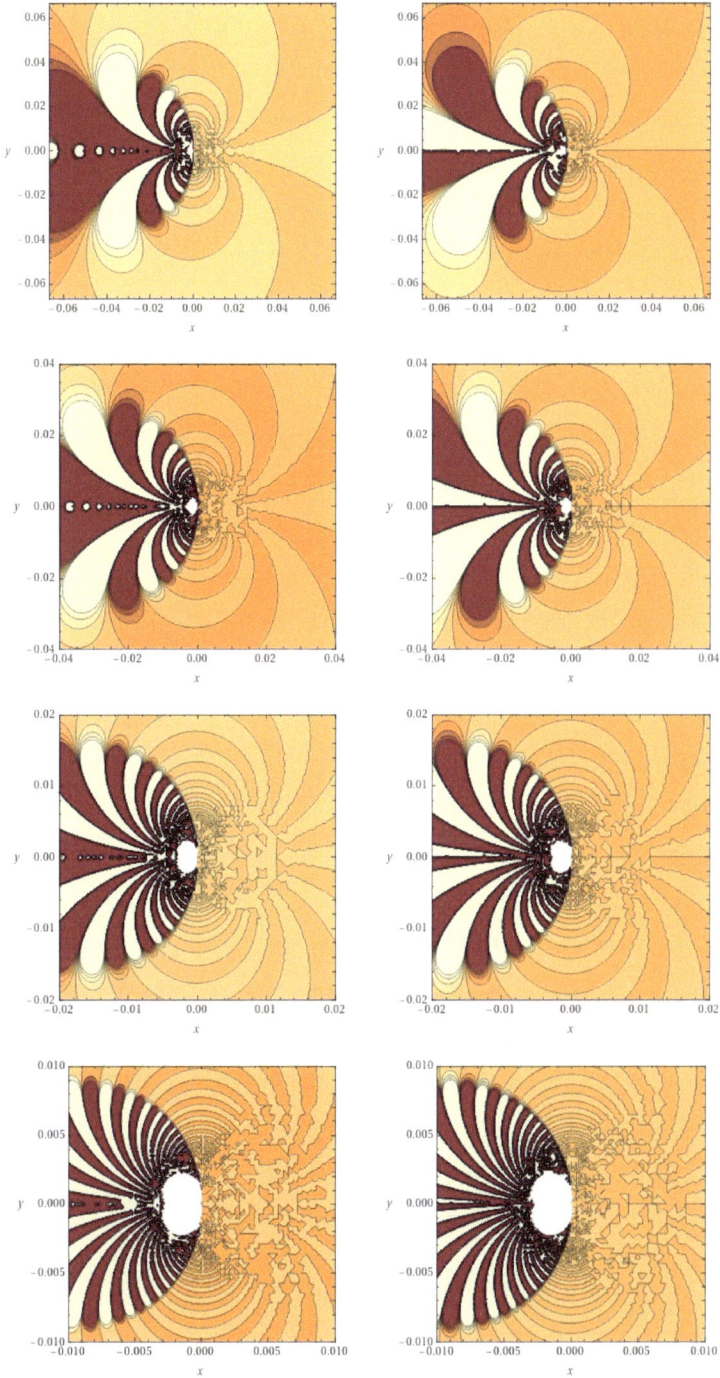

Real (left) and imaginary (right) plots of the inner action of balance 2 over a range of scales, under inverse-complex argument (continued).

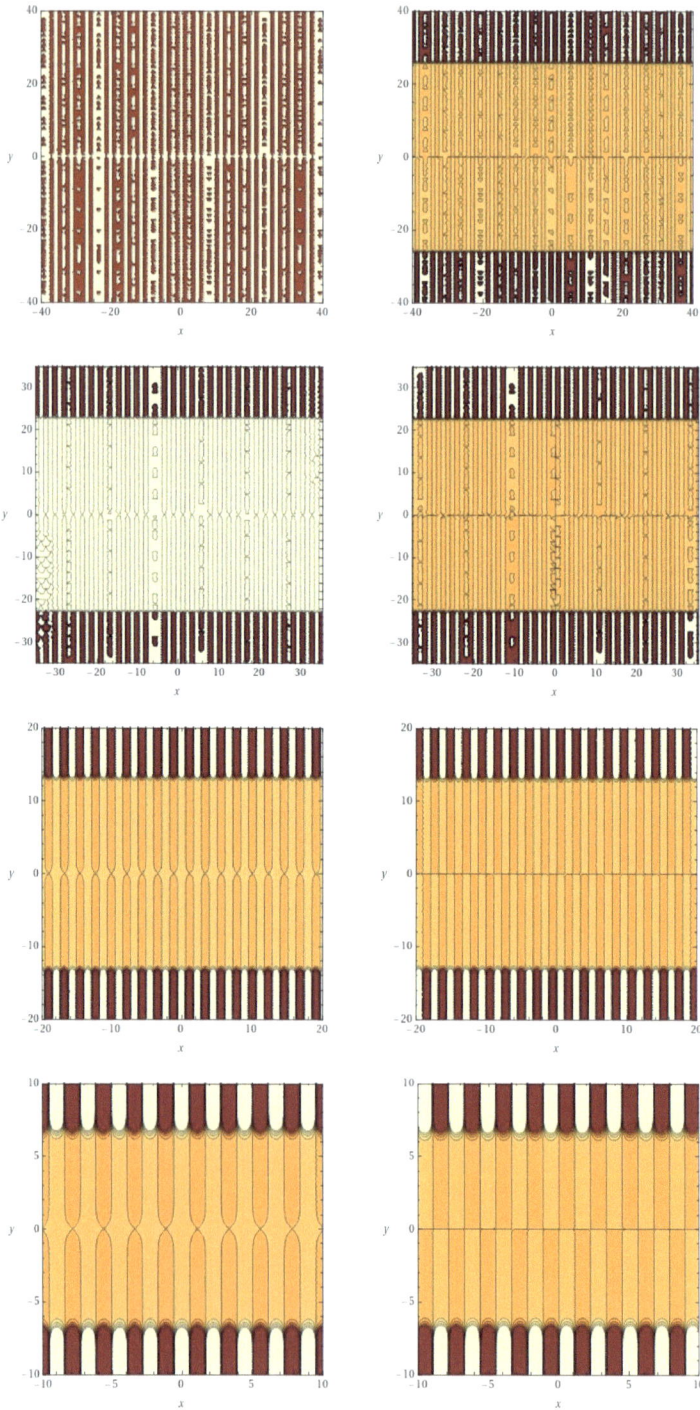

Real (left) and imaginary (right) plots of the inner action of balance 3 over a range of scales, under complex argument.

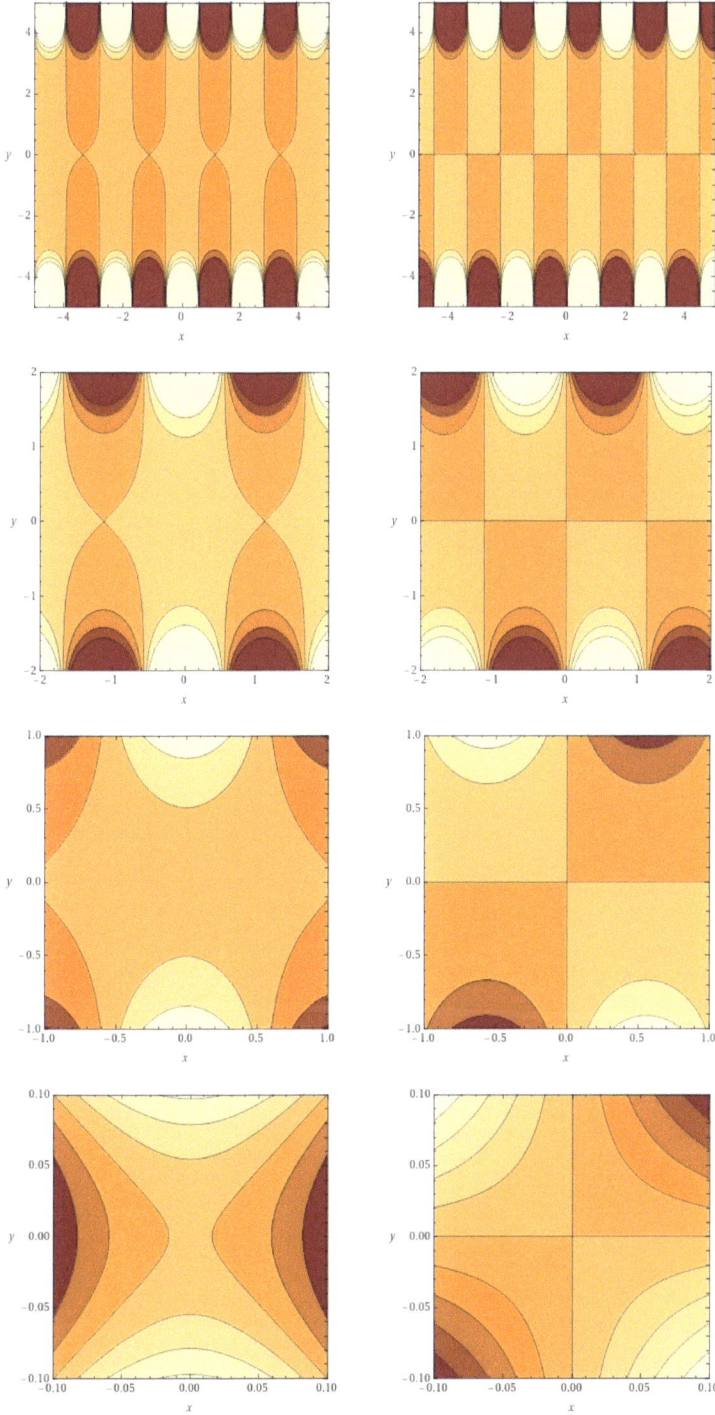

Real (left) and imaginary (right) plots of the inner action of balance 3 over a range of scales, under complex argument (continued).

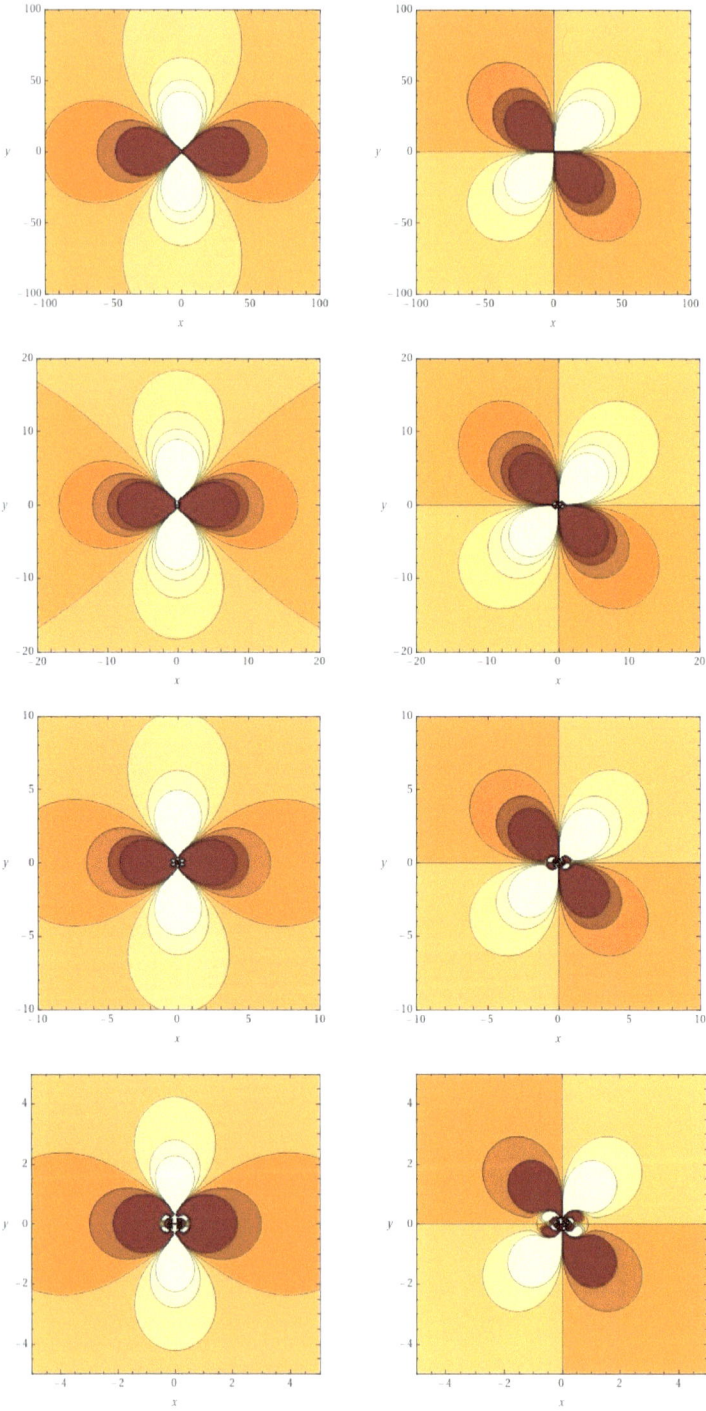

Real (left) and imaginary (right) plots of the inner action of balance 3 over a range of scales, under inverse-complex argument.

144

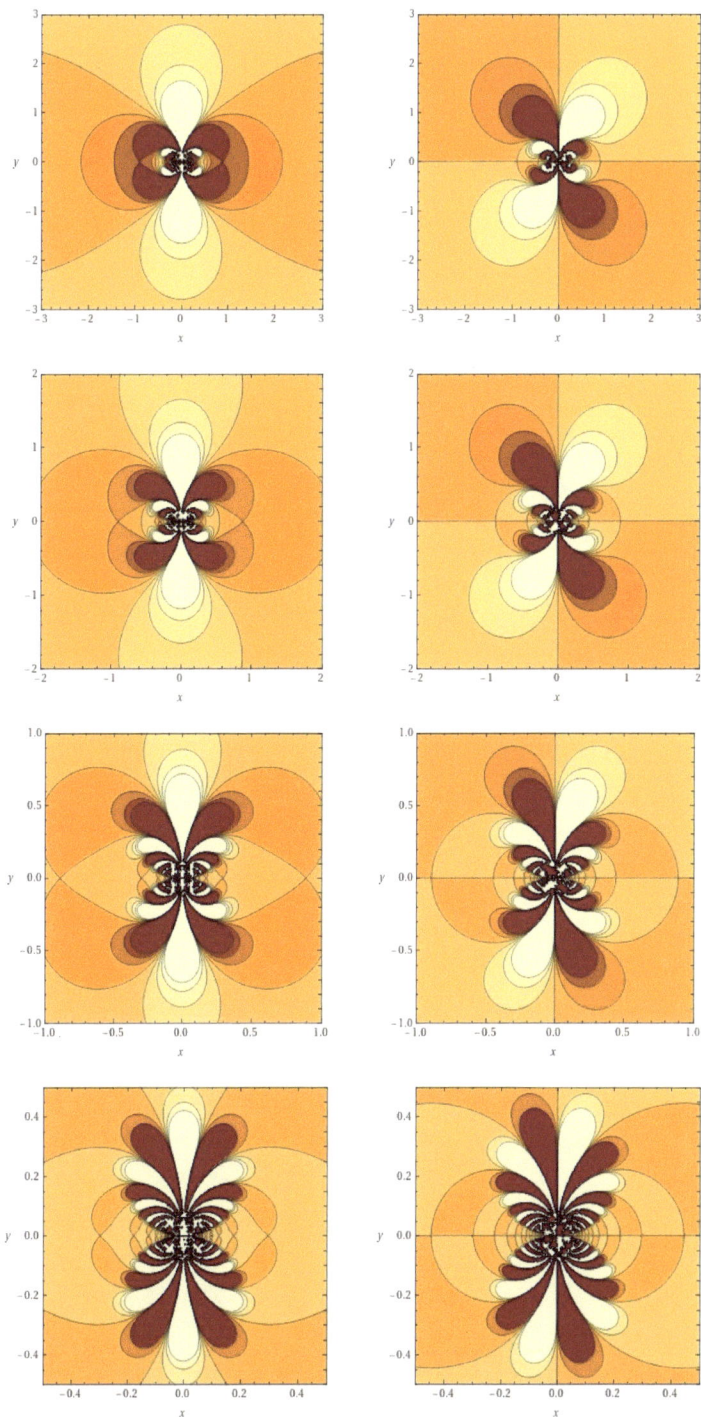

Real (left) and imaginary (right) plots of the inner action of balance 3 over a range of scales, under inverse-complex argument (continued).

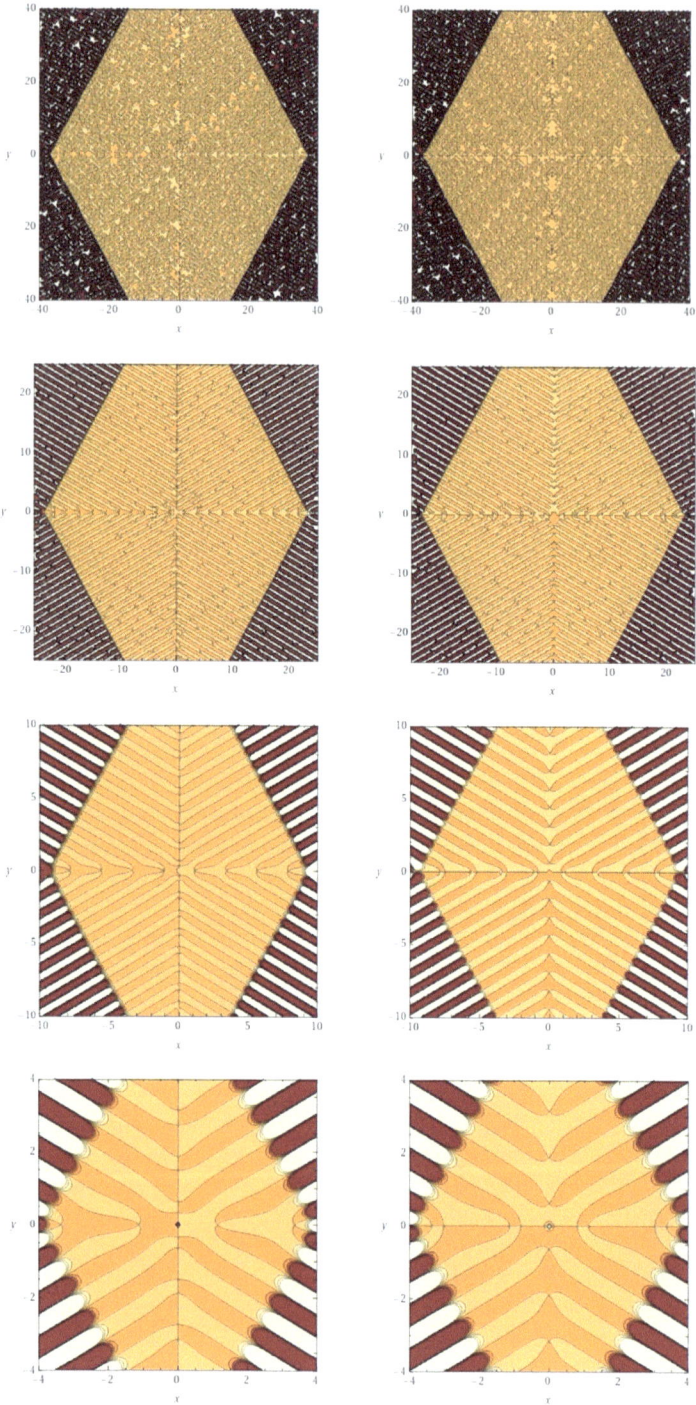

Real (left) and imaginary (right) plots of the inner action of balance 4 over a range of scales, under complex argument.

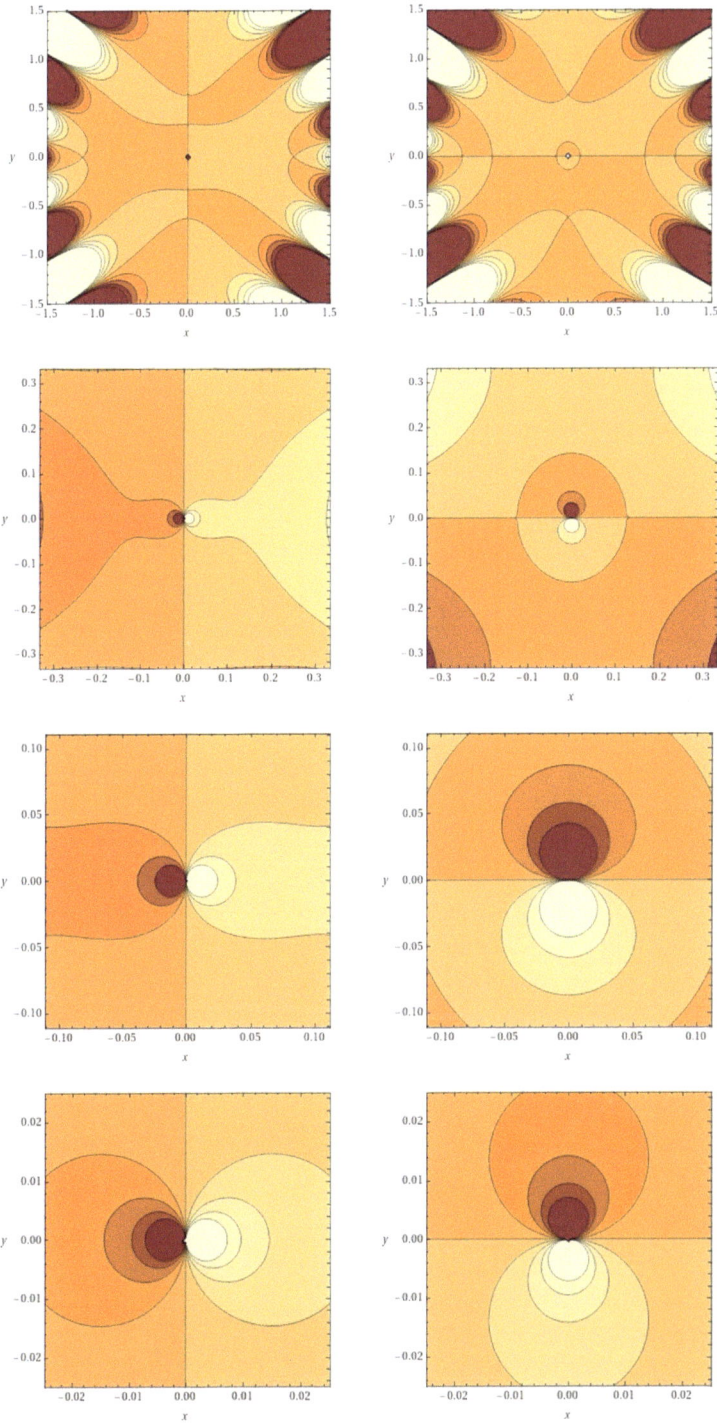

Real (left) and imaginary (right) plots of the inner action of balance 4 over a range of scales, under complex argument (continued).

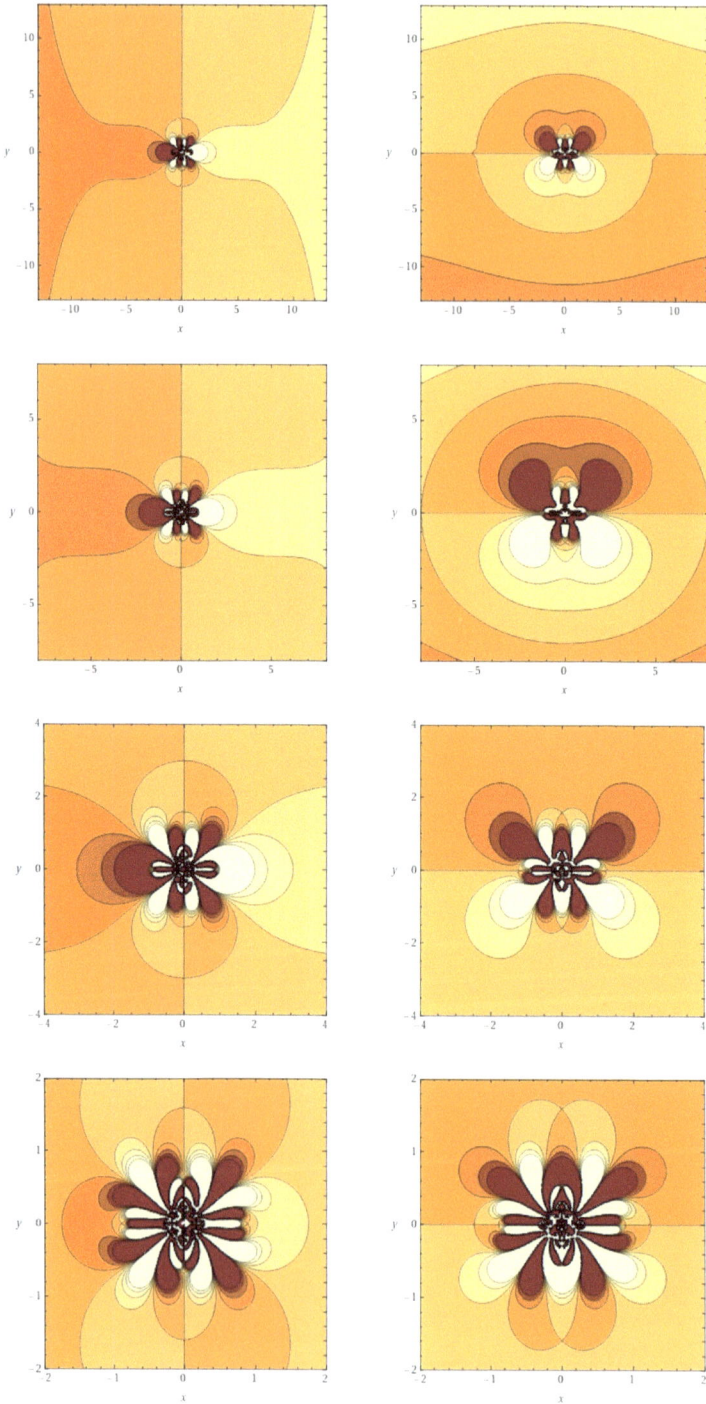

Real (left) and imaginary (right) plots of the inner action of balance 4 over a range of scales, under inverse-complex argument.

148

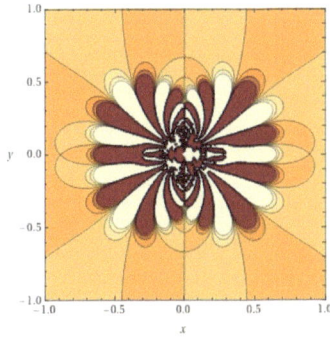

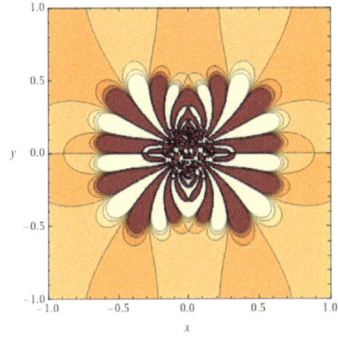

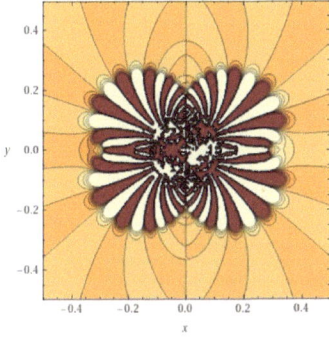

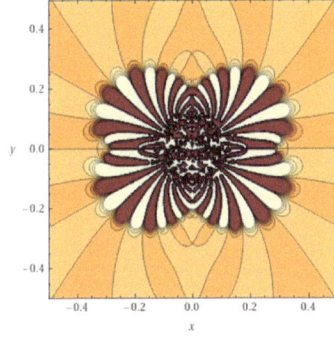

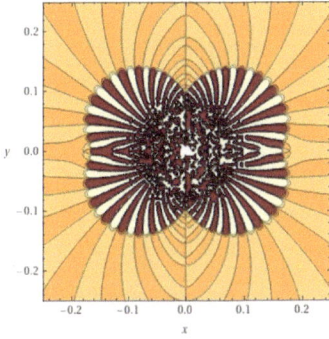

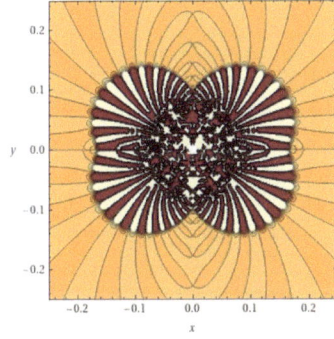

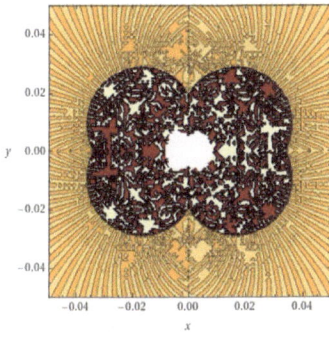

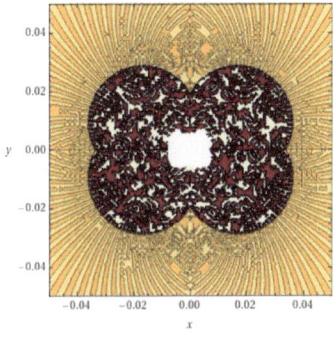

Real (left) and imaginary (right) plots of the inner action of balance 4 over a range of scales, under inverse-complex argument (continued).

Appendix b: charge and mass divisions (equations only)

$$\frac{1}{\text{ж}} + \text{ж} + \frac{\text{ж}^3}{2\pi} = \left(i^i \right)^{-\frac{\pi}{2}} - m_p \qquad \text{the hyperbolic vortex equation}$$

$$\text{ж}_1 \text{ж}_2 \text{ж}_3 \text{ж}_4 = 2\pi \qquad \text{product}$$

$$\text{ж}_1 + \text{ж}_2 + \text{ж}_3 + \text{ж}_4 = 0 \qquad \text{sum}$$

$$\text{ж}_1 + \text{ж}_2 + Re(\text{ж}_3) + Re(\text{ж}_4) = 0 \qquad \text{real sum}$$

$$Im(\text{ж}_3) + Im(\text{ж}_4) = 0 \qquad \text{imaginary sum}$$

$$\text{ж}_1{}^2 + \text{ж}_2{}^2 + \text{ж}_3{}^2 + \text{ж}_4{}^2 = -4\pi \qquad \text{square sum}$$

$$Re(\text{ж}_3) = Re(\text{ж}_4) = -\frac{1}{2}(\text{ж}_1 + \text{ж}_2) \qquad \text{– joined twins}$$

$$Im(\text{ж}_3) = -Im(\text{ж}_4) = \frac{1}{2}(Im(\text{ж}_3) - Im(\text{ж}_4)) \qquad \text{+ split twins}$$

$$\text{ж}_3 + \text{ж}_4 = 2\,Re(\text{ж}_3) = 2\,Re(\text{ж}_4) \qquad \text{partial sum}$$

$$\text{ж}_3 - \text{ж}_4 = 2\,Im(\text{ж}_3) = 2\,Im(\text{ж}_4) \qquad \text{partial difference}$$

$$Re(\text{ж}_3)^2 + Re(\text{ж}_4)^2 = \frac{1}{2}(\text{ж}_1 + \text{ж}_2)^2 \qquad \text{real square sum}$$

$$Re(\text{ж}_3)^2 + Im(\text{ж}_3)^2 = Re(\text{ж}_4)^2 + Im(\text{ж}_4)^2 = \text{ж}_3 \text{ж}_4 = \text{ж}_r{}^2 \qquad \text{pair}$$

$$Re(\text{ж}_3)^2 - Re(\text{ж}_4)^2 = Im(\text{ж}_3)^2 - Im(\text{ж}_4)^2 = \text{ж}_3 \text{ж}_4 = \text{ж}_r{}^2 \qquad \text{pair}$$

$$Im(\text{ж}_3)^2 + Im(\text{ж}_4)^2 = Im(\text{ж}_3)^4 = Im(\text{ж}_4)^4 \qquad \text{imaginary square sum}$$

$$\left(\frac{\text{ж}_1{}^2 + \text{ж}_2{}^2 + \text{ж}_3{}^2 + \text{ж}_4{}^2}{\text{ж}_1 \times \text{ж}_2 \times \text{ж}_3 \times \text{ж}_4} \right) = -2 \qquad \text{self intersection number}$$

$$\text{ж}_1{}^2 + \text{ж}_2{}^2 + \text{ж}_3{}^2 + \text{ж}_4{}^2 = -\left(2\,\Gamma\left(\tfrac{1}{2}\right) \right)^2 \qquad \text{square sum gamma split}$$

$$\sum_{k=1}^{4} Ж_k = 0 \qquad\qquad \prod_{k=1}^{4} Ж_k = 2\pi$$

$$\sum_{k=1}^{4} Ж_k{}^2 = -4\pi \qquad\qquad \prod_{k=1}^{4} Ж_k{}^2 = 4\pi^2$$

$$-\pi \sum_{k=1}^{4} Ж_k{}^2 = \prod_{k=1}^{4} Ж_k{}^2 \qquad \text{square sum to square product}$$

$$-\frac{1}{2}\sum_{k=1}^{4} Ж_k{}^2 = \prod_{k=1}^{4} Ж_k \qquad \text{half square sum = product}$$

$$Ж_3 = Ж_r \, e^{\,Ж_\theta i} \qquad\qquad Ж_3 Ж_4 = Ж_r{}^2$$

$$Ж_4 = Ж_r \, e^{-Ж_\theta i} \qquad\qquad Ж_1 Ж_2 Ж_r{}^2 = 2\pi$$

$$Ж_r = \sqrt{Ж_3 Ж_4} \qquad\qquad \log\left(\frac{Ж_3}{Ж_r}\right) = Ж_\theta i$$

$$Ж_1 + Ж_2 + \frac{Ж_r{}^2}{Ж_3} + \frac{Ж_r{}^2}{Ж_4} = 0 \qquad\qquad \log\left(\frac{Ж_4}{Ж_r}\right) = -Ж_\theta i$$

$$Ж_\theta = \tan^{-1}\left(\frac{Re(Ж_3)}{Im(Ж_4)}\right) + \frac{\pi}{2}$$

$$m_e = 2V_{fe}\, m_p{}^4 \left(1 + \left(sinh \left(sinh \left(\frac{!\,n}{b} \right) \right)^{-1} \right)^{-1} \boxplus \right)$$

$$\left(\frac{m_H - m_Z}{m_W} \right) = \left(\frac{1}{3} \right)^2 (\mu + 3 + \pi)(2^{-1})\left(1 + 6 \left(\frac{\text{Ж}_r}{W(1)} \right)^4 \boxplus \right)$$

$$\left(\frac{m_N - m_+}{m_e} \right) = \left(\frac{1}{3} \right)(\mu + 3 + \pi)(2^0)\left(1 + \left(\frac{2}{3} \right) e^{3\gamma} \boxplus \right)$$

$$\left(\frac{m_d - m_u}{m_e} \right) = \left(\frac{1}{3} \right)(\mu + 3 + \pi)(2^{+1})\left(1 - \left(\frac{b}{n} \right)\sqrt{S_S}\; \frac{\text{Ж}_r{}^4}{4} \boxplus \right)$$

$$\left(\frac{m_c - 2m_s}{m_\mu} \right) = \left(\frac{1}{3} \right)(\mu + 3 + \pi)(2^2)\left(1 - \left(\frac{3}{V_h^*} \right)^3 \frac{\text{Ж}_r{}^4}{2} \boxplus \right)$$

$$\left(\frac{m_t - m_\tau}{m_b} \right) = \left(\frac{1}{3} \right)(\mu + 3 + \pi)\left(2^{2^2} \right)\left(1 + \left(\frac{b}{6} \right) D_{DHA}{}^2\, \text{Ж}_r{}^2 \boxplus \right)$$

$$\left(\frac{m_{\nu_\tau} - m_{\nu_\mu}}{m_{\nu_e}} \right) = \left(\frac{1}{3} \right)(\mu + 3 + \pi)\left(2^{2^{2^2}} \right)\left(1 - ??? \boxplus \right)$$

$$\frac{m_b + m_c + m_t}{\left(\sqrt{m_b} + \sqrt{m_c} + \sqrt{m_t} \right)^2} = \left(\frac{2}{3} \right)^3 (\alpha_F - 1)^2 \left(1 - \sqrt{W_{We}}\, \left(2\, Im(\text{ж}_3) \right)^4 \boxplus \right)$$

$$\left(\frac{m_e}{m_+} \right)\left(\frac{\text{Ж}_2}{\text{Ж}_1} \right)^2 = \left(\frac{2}{3} \right)^2 (\alpha_F - 1)^2 \left(1 - \left(\frac{1}{3} \right) e^{3\gamma} \boxplus \right)$$

$$\frac{m_e + m_\mu + m_\tau}{\left(\sqrt{m_e} + \sqrt{m_\mu} + \sqrt{m_\tau} \right)^2} = \left(\frac{2}{3} \right)^1 (\alpha_F - 1)^0 \left(1 + 2\, L^2 \left(Re(\text{ж}_3) \right)^3 \boxplus \right)$$

$$\frac{m_H + m_Z + m_W}{\left(\sqrt{m_H} + \sqrt{m_Z} + \sqrt{m_W} \right)^2} = \left(\frac{2}{3} \right)^{-\frac{1}{2}} (\alpha_F - 1)^{-e^{2\gamma}} \left(1 + \left(\frac{3}{n} \right)\sqrt{P_{up}{}^n}\; \text{Ж}_r{}^2 \boxplus \right)$$

$$\frac{m_u + m_s + m_d}{\left(\sqrt{m_u} + \sqrt{m_s} + \sqrt{m_d} \right)^2} = \left(\frac{2}{3} \right)^0 \gamma\,(\alpha_F - 1)^0 \left(1 - 6 \left(2V_{fe} \right) \text{Ж}_r{}^3 \boxplus \right)$$

$$\frac{m_{\nu_e} + m_{\nu_\mu} + m_{\nu_\tau}}{\left(\sqrt{m_{\nu_e}} + \sqrt{m_{\nu_\mu}} + \sqrt{m_{\nu_\tau}} \right)^2} = \left(\frac{2}{3} \right)^{-\frac{1}{2}} (\alpha_F - 1)^{+e^{\pi i}} \left(1 + ??? \boxplus \right)$$

$$\left(\frac{m_e}{m_\mu}\right) = \frac{\text{Ж}_1{}^2}{\sinh(C_{CFP})}\left(1 + \left(\frac{d_0\, d_1}{4\pi}\right)\boxplus\right)$$

$$\frac{d_0}{d_3}\left(\frac{m_e}{m_b}\right) = 4\pi\,\text{Ж}_1{}^4 \qquad\qquad \frac{d_0}{n}\left(\frac{m_e}{m_\tau}\right) = 2V_{fe}\,\text{Ж}_1{}^3$$

$$\frac{1}{d_3}\left(\frac{m_N}{m_c}\right) = 4\pi\,\text{Ж}_1{}^2 \qquad\qquad d_1 d_2\left(\frac{m_+}{m_Z}\right) = 2\,\varphi\,\text{Ж}_\theta$$

$$\frac{3\,d_0}{2\,d_3}\left(\frac{m_W}{m_b}\right) = (4\pi)^2 \qquad\qquad \left(\frac{m_\tau}{m_u}\right) = \left(2\pi\,\text{Ж}_r\right)^2$$

$$\frac{d_2}{d_1 d_3}\left(\frac{m_H}{m_c}\right) = 2\pi \qquad\qquad \frac{d_1}{d_2}\left(\frac{m_N}{m_H}\right) = 2\,\text{Ж}_1{}^2$$

$$\frac{2}{3}\left(\frac{m_e}{m_W}\right) = \frac{\text{Ж}_1{}^4}{4\pi} \qquad\qquad d_2{}^2\left(\frac{m_N}{m_t}\right) = \frac{(4\pi)^2}{\text{Ж}_r{}^3}$$

$$\frac{1}{\sqrt{3n}}\left(\frac{m_+}{m_s}\right) = P_{up} \qquad\qquad \left(\frac{m_\mu}{m_b}\right) = 6\,\text{Ж}_1{}^2\,\bar{s}_{lse}{}^4$$

$$m_{\nu_e} = ? \qquad\qquad\qquad \left(\frac{m_c}{m_d}\right) = \sqrt{\frac{\bar{s}_{lse}}{2}}\,\text{Ж}_r{}^4$$

$$m_{\nu_\mu} = ?$$

$$m_{\nu_\tau} = ?$$

Where m_e, m_+, m_N, m_μ, m_τ, m_Z, m_W, m_H, m_t, m_c, m_b, m_s, m_d, m_u, m_{ν_τ}, m_{ν_μ} and m_{ν_e} are respectively the masses of the: electron, proton, neutron, muon, tau, Z boson, W boson, Higgs boson, truth (top) quark, charm quark, beauty (bottom) quark, strange quark, down quark, up quark, tau neutrino, muon neutrino, and the electron neutrino, π = Archimedes' constant, V_{fe} = the volume complement of the hyperbolic figure eight knot, Ж_1 = the 1st hyperbolic vortex partition constant, Ж_r = the hyperbolic vortex radius constant, P_{up} = the universal parabolic constant, $n = 5$, φ = the golden ratio, C_{CFP} = the fixed point of the hyperbolic cotangent, $\bar{s}_{lse}$ = the mean line-between-square edges length (in hypercube line picking), d_0 = the 44 derangements of the time boundary, d_1 = the 35 derangements of the space boundary, d_2 = the 18 derangements of the charge boundary, and d_3 = the 8 derangements of the mass boundary.

154

Appendix c: constants of Nature (equations only)

$$\alpha = \text{Ж}_1{}^2$$

$$e = \text{Ж}_1 q_p$$

$$\alpha_G = \left(\frac{m_e}{m_p}\right)^2$$

$$\mu_0 = 4\pi \; \boxplus \left(1 + 4\pi \left(\frac{\pi}{2}\right)^{-1} \boxplus \right)$$

$$S_{mi} = \frac{1}{\text{Ж}_1}\left(\frac{m_e{}^2}{t_p\, q_p\, m_p}\right)\left(1 - \left(3\, Im\left(i^{i^{i^{\cdots}}}\right)\right)^2 2\pi \; \boxplus \right)$$

$$R_K = \frac{2\pi}{\text{Ж}_1{}^2}\left(\frac{l_p{}^2\, m_p}{t_p\, q_p{}^2}\right)\left(1 + \frac{1}{2}Re\left(i^{i^{i^{\cdots}}}\right)\text{Ж}_r{}^2 \; \boxplus \right)$$

$$H_C = \frac{\text{Ж}_1{}^2}{2\pi}\left(\frac{t_p\, q_p{}^2}{l_p{}^2 m_p}\right)\left(1 - \frac{1}{2}Re\left(i^{i^{i^{\cdots}}}\right)\text{Ж}_r{}^2 \; \boxplus \right)$$

$$\Phi_0 = \frac{\pi}{\text{Ж}_1}\left(\frac{l_p{}^2\, m_p}{t_p\, q_p}\right)\left(1 + sec^2\left(\left(\frac{1}{2}\right)^2\right)\text{Ж}_r \; \boxplus \right)$$

$$K_J = \frac{\text{Ж}_1}{\pi}\left(\frac{t_p\, q_p}{l_p{}^2\, m_p}\right)\left(1 - sec^2\left(\left(\frac{1}{2}\right)^2\right)\text{Ж}_r \; \boxplus \right)$$

$$q_c = \pi\left(\frac{l_p{}^2\, m_p}{t_p\, m_e}\right)\left(1 + \sqrt{\zeta(2)}\;\text{Ж}_\theta{}^2 \; \boxplus \right)$$

$$G_0 = \frac{\text{Ж}_1{}^2}{\pi}\left(\frac{t_p\, q_p{}^2}{l_p{}^2 m_p}\right)\left(1 - \sqrt{\zeta(3)}\;\text{Ж}_\theta{}^2 \; \boxplus \right)$$

$$c_2 = 2\pi\left(l_p\, T_p\right)\left(1 - sinh^2(C_{CFP})\,\text{Ж}_r{}^2 \; \boxplus \right)$$

$$\hbar = \left(\frac{l_p{}^2\, m_p}{t_p}\right)\left(1 + \left(\frac{4\pi}{sinh^2(2)}\right)\text{Ж}_r \; \boxplus \right)$$

$$\kappa = \left(\frac{l_p{}^3\, m_p}{t_p{}^2\, q_p{}^2} \right) \left(1 + \left(\frac{\pi}{d_2} \right) \text{ж}_r \ \boxplus \right)$$

$$\varepsilon_0 = \frac{1}{4\pi} \left(\frac{t_p{}^2\, q_p{}^2}{l_p{}^3\, m_p} \right) \left(1 - \left(\frac{\pi}{d_2} \right) \text{ж}_r \ \boxplus \right)$$

$$c = \left(\frac{l_p}{t_p} \right) \left(1 - \left(\frac{2\pi}{d_1} \right) \text{ж}_r{}^2 \ \boxplus \right)$$

$$F = N_A\, \text{ж}_1\, q_p \left(1 + \left(\frac{n}{d_0} \right) \frac{\text{ж}_r{}^2}{2\pi} \ \boxplus \right)$$

$$g_N = -\frac{\Gamma(n)}{2\pi} \left(\frac{m_N}{m_+} \right) \left(1 + n \left(\frac{d_2}{4\pi} \right) \text{ж}_r{}^4 \ \boxplus \right)$$

$$G = \left(\frac{l_p{}^3}{t_p{}^2\, m_p} \right) \left(1 - \left(\frac{\pi}{2} \right) \Gamma(n)\, \text{ж}_r{}^2 \ \boxplus \right)$$

$$k_B = \left(\frac{l_p{}^2\, m_p}{t_p{}^2\, T_p} \right) \left(1 - \left(\frac{\pi}{2} \right)^2 \text{ж}_r{}^2 \ \boxplus \right)$$

$$\omega_C = \left(\frac{m_e}{t_p\, m_p} \right) \left(1 - \frac{1}{2}\Gamma(n) \ \boxplus \right)$$

$$c_{1L} = 4\pi \left(\frac{l_p{}^4\, m_p}{t_p{}^3} \right) \left(1 - \left(\frac{d_3}{d_1} \right) 4\pi \ \boxplus \right)$$

$$c_1 = (2\pi)^2 \left(\frac{l_p{}^4\, m_p}{t_p{}^3} \right) \left(1 - \left(\frac{d_3}{d_1} \right) V_{fe}\, \text{ж}_r \ \boxplus \right)$$

$$a_0 = \frac{1}{\text{ж}_1{}^2} \left(\frac{l_p\, m_p}{m_e} \right) \left(1 + \left(\frac{d_1}{d_3} \right) \text{ж}_\theta \ \boxplus \right)$$

$$r_e = \text{ж}_1{}^2 \left(\frac{l_p\, m_p}{m_e} \right) \left(1 + \left(\frac{d_1}{d_3} \right) 2 \ \boxplus \right)$$

$$Z_0 = 4\pi \left(\frac{l_p{}^2 m_p}{t_p\, q_p{}^2} \right) \left(1 + \left(\frac{d_1}{2} \right) \frac{\sqrt{G_{Gi}}}{\text{ж}_\theta{}^2} \ \boxplus \right)$$

$$E_h = ж_1{}^4 \left(\frac{l_p{}^2 \, m_e}{t_p{}^2} \right) \left(1 - \left(\frac{b}{n} \right) \frac{e^\pi}{ж_r} \boxplus \right)$$

$$\lambda_C = 2\pi \left(\frac{l_p \, m_p}{m_e} \right) \left(1 + \left(\frac{b}{2} \right) \alpha_F \boxplus \right)$$

$$\mu_B = \frac{ж_1}{2} \left(\frac{l_p{}^2 \, q_p \, m_p}{t_p \, m_e} \right) \left(1 + \left(\frac{2}{3} \right) V_{fe} \, ж_\theta{}^2 \boxplus \right)$$

$$\mu_N = \frac{ж_1}{2} \left(\frac{l_p{}^2 \, q_p \, m_p}{t_p \, m_+} \right) \left(1 + \frac{1}{n} \left(\frac{2}{3} \right) V_{fe} \, ж_r{}^2 \boxplus \right)$$

$$N_A = \frac{6 \, ж_1{}^2}{e^\gamma} \left(\frac{1}{q_p \, m_p} \right) \left(1 - \left(2 \, ж_2 \, P_{up} \right)^2 \boxplus \right)$$

$$\sigma = \frac{\zeta(2)}{2n} \left(\frac{m_p}{t_p{}^3 \, T_p{}^4} \right) \left(1 + \left(\frac{P_{up}}{2 \, ж_1} \right)^2 \boxplus \right)$$

$$R = \frac{6 \, ж_1{}^2}{e^\gamma} \left(\frac{l_p{}^2}{t_p{}^2 \, q_p \, T_p} \right) \left(1 - \sqrt{m_R} \, Im(ж_3)^4 \boxplus \right)$$

$$\gamma_+ = \left(2 \, ж_r{}^2 P_{up} \right)^2 \left(\frac{t_p}{q_p \, m_e} \right) \left(1 + \frac{1}{d_3} \left(3 \, Im \left(i^{i^{i^{\cdot}}} \right) \right) ж_r{}^3 \boxplus \right)$$

$$m_u = \left(\frac{1}{2n} \right)^3 \frac{e^\gamma}{6 \, ж_1{}^2} \left(q_p \, m_p \right) \left(1 + d_2{}^2 \left(2 \, Re \left(i^{i^{i^{\cdot}}} \right) \right) \boxplus \right)$$

$$\sigma_e = 4\pi \, ж_1{}^4 \left(\frac{2}{3} \right) \left(\frac{l_p \, m_p}{m_e} \right)^2 \left(1 + sin \left(\frac{b\pi}{2 \, \Gamma(n)} \right) ж_2{}^3 \boxplus \right)$$

$$R_\infty = \frac{ж_1{}^4}{4\pi} \left(\frac{m_e}{l_p \, m_p} \right) \left(1 - \left(\frac{2 \, \Gamma(n)}{d_0} \right) ж_\theta{}^3 \boxplus \right)$$

$$g_\mu = -\sqrt{b} \, C_{R1}{}^2 \, ж_1 \left(\frac{m_N}{m_\mu} \right) \left(1 + sec \left(cot \left(\frac{2 \, \Gamma(n)}{b} \right) \right) \boxplus \right)$$

$$g_e = -\sqrt{b}\, C_{R1}{}^2\, \text{Ж}_1 \left(\frac{m_N}{m_\mu} \right) \left(1 - sec\left(cot\left(\frac{2n}{d_0} \right) \right) \text{Ж}_\theta{}^3 \;\boxplus\; \right)$$

$$g_+ = 3\sqrt{3\, G_g} \left(\frac{m_+}{m_N} \right) \left(1 + e^{-1/e}\, \text{Ж}_r{}^3 \;\boxplus\; \right)$$

$$N_\mu = -\frac{\sqrt{2}\, G_g}{\text{Ж}_r{}^2} \left(\frac{l_p{}^2 q_p\, m_p}{t_p\, m_+} \right) \left(1 - \left(\frac{4\pi}{17} \right) \boxplus \right)$$

$$b_{entropy} = L_{LL}{}^{\frac{1}{4}} \left(\frac{b}{n} \right)^{\frac{1}{2}} \left(l_p T_p \right) \left(1 - \frac{1}{\sqrt{2}}\; \boxplus \right)$$

$$\frac{r_N}{r_e} = \left(\frac{4\pi}{\Gamma(n)} \right)^2$$

$$\frac{r_N}{r_+} = -\zeta'(0)$$

$$C_{R1} = \cfrac{1}{1 + \cfrac{e^{-2\pi}}{1 + \cfrac{e^{-4\pi}}{1 + \cfrac{e^{-6\pi}}{1 + \cdots}}}}$$

$$\frac{em\; force}{strong\; force} = \text{Ж}_1{}^2$$

$$\frac{r_e}{a_0} = \text{Ж}_1{}^4$$

$$He^+ = \frac{a_0}{2}$$

$$Li^{2+} = \frac{a_0}{3}$$

$$\psi(r) = \left(\pi a_0{}^3 \right)^{-1/2} e^{-r/a_0} = \frac{a_0{}^{-3/2}}{\Gamma\left(\frac{1}{2} \right)}\, e^{-r/a_0}$$

Appendix d: geometric identities

$$\frac{1}{1-x} = x^0 + x^1 + x^2 + x^3 + x^4 + x^5 + x^6 + x^7 + x^8 + x^9 + \cdots$$

$$\frac{1}{1+x} = x^0 - x^1 + x^2 - x^3 + x^4 - x^5 + x^6 - x^7 + x^8 - x^9 - \cdots$$

$$1^2 + 2^2 + 3^2 + 4^2 + 5^2 + 6^2 + 7^2 + \cdots + k^2 = \frac{k(k-1)(2k+1)}{6}$$

$$1^3 + 2^3 + 3^3 + 4^3 + 5^3 + \cdots + k^3 = \left(1 + 2 + 3 + 4 + \cdots + k \right)^2$$

$$e = \frac{1}{0!} + \frac{1}{1!} + \frac{1}{2!} + \frac{1}{3!} + \frac{1}{4!} + \frac{1}{5!} + \frac{1}{6!} + \frac{1}{7!} + \frac{1}{8!} + \frac{1}{9!} + \frac{1}{10!} + \frac{1}{11!} + \cdots$$

$$\frac{1}{e} = \frac{1}{0!} - \frac{1}{1!} + \frac{1}{2!} - \frac{1}{3!} + \frac{1}{4!} - \frac{1}{5!} + \frac{1}{6!} - \frac{1}{7!} + \frac{1}{8!} - \frac{1}{9!} + \frac{1}{10!} - \frac{1}{11!} + \cdots$$

$$e^x = \frac{x^0}{0!} + \frac{x^1}{1!} + \frac{x^2}{2!} + \frac{x^3}{3!} + \frac{x^4}{4!} + \frac{x^5}{5!} + \frac{x^6}{6!} + \frac{x^7}{7!} + \frac{x^8}{8!} + \frac{x^9}{9!} + \frac{x^{10}}{10!} + \cdots$$

$$cosh(x) = \frac{x^0}{0!} + \frac{x^2}{2!} + \frac{x^4}{4!} + \frac{x^6}{6!} + \frac{x^8}{8!} + \frac{x^{10}}{10!} + \frac{x^{12}}{12!} + \frac{x^{14}}{14!} + \frac{x^{16}}{16!} + \cdots$$

$$cos(x) = \frac{x^0}{0!} - \frac{x^2}{2!} + \frac{x^4}{4!} - \frac{x^6}{6!} + \frac{x^8}{8!} - \frac{x^{10}}{10!} + \frac{x^{12}}{12!} - \frac{x^{14}}{14!} + \frac{x^{16}}{16!} - \cdots$$

$$sinh(x) = \frac{x^1}{1!} + \frac{x^3}{3!} + \frac{x^5}{5!} + \frac{x^7}{7!} + \frac{x^9}{9!} + \frac{x^{11}}{11!} + \frac{x^{13}}{13!} + \frac{x^{15}}{15!} + \frac{x^{17}}{17!} + \cdots$$

$$sin(x) = \frac{x^1}{1!} - \frac{x^3}{3!} + \frac{x^5}{5!} - \frac{x^7}{7!} + \frac{x^9}{9!} - \frac{x^{11}}{11!} + \frac{x^{13}}{13!} - \frac{x^{15}}{15!} + \frac{x^{17}}{17!} - \cdots$$

$$\tan^{-1}(x) = \frac{x^1}{1} - \frac{x^3}{3} + \frac{x^5}{5} - \frac{x^7}{7} + \frac{x^9}{9} - \frac{x^{11}}{11} + \frac{x^{13}}{13} - \frac{x^{15}}{15} - \cdots$$

$$(a + bi) = re^{i\theta} \qquad\qquad \text{complex conversion}$$

$$(a+b)^n = \sum_{k=0}^{n} \binom{n}{k} a^k b^{n-k} \qquad \text{binomial factorization}$$

$$\binom{n}{k} = \frac{n!}{k!\,(n-k)!} \qquad\qquad \int_0^1 x^n \, dx = \frac{1}{n+1}$$

$$(1+x)^p = \frac{1}{0!} + p\frac{x^1}{1!} + p(p-1)\frac{x^2}{2!} + p(p-1)(p-2)\frac{x^3}{3!} + \cdots$$

$$\left(\frac{2}{1}\right)^{1/1} \left(\frac{2^2}{3}\right)^{1/2} \left(\frac{2^3\,4}{3^3}\right)^{1/3} \left(\frac{2^4\,4^4}{3^6\,5}\right)^{1/4} \cdots = e$$

$$\left(\frac{2}{1}\right)^{1/2} \left(\frac{2^2}{3}\right)^{1/3} \left(\frac{2^3\,4}{3^3}\right)^{1/4} \left(\frac{2^4\,4^4}{3^6\,5}\right)^{1/5} \cdots = e^\gamma$$

$$\left(\frac{2}{1}\right)^{1/2} \left(\frac{2^2}{3}\right)^{1/4} \left(\frac{2^3\,4}{3^3}\right)^{1/8} \left(\frac{2^4\,4^4}{3^6\,5}\right)^{1/16} \cdots = \frac{\pi}{2}$$

$$V_{fe} = \frac{3}{2}\sqrt{3}\left(1 - \sum_{k=0}^{\infty} \frac{1}{(3k+2)^2} + \sum_{n=0}^{\infty} \frac{1}{(3k+1)^2}\right)$$

$$V_{fe} = \frac{1}{2}\left(1 + \cfrac{1}{1 + \cfrac{1}{2 + \cfrac{1}{1 + \cfrac{1}{2 + \cfrac{1}{1 + \cfrac{1}{2+\ddots}}}}}}\right) 3\left(1 + \frac{1}{1^2} - \frac{1}{2^2} + \frac{1}{4^2} - \frac{1}{5^2} + \frac{1}{7^2} - \cdots\right)$$

$$\prod_{k=0}^{\infty} \left(1 - q^k\right)^{k/5} = \cfrac{q^0}{1 + \cfrac{q^1}{1 + \cfrac{q^2}{1 + \cfrac{q^3}{1 + \cfrac{q^4}{1+\ddots}}}}} \qquad \text{Rogers} - \text{Ramanujan identities}$$

fixed points of the hyperbolic and circular functions

$cos(x) = x$ D_{Do} the Dottie number

$coth(x) = x$ C_{CFP} Real fixed point of the hyperbolic cotangent

$sinh(x) = x$ 0

$sinh^{-1}(x) = x$ 0

$tanh(x) = x$ 0

$tanh^{-1}(x) = x$ 0

$sin(x) = x$ 0

$sin^{-1}(x) = x$ 0

$$sinh(C_{CFP}) = \frac{1}{L_{LL}}$$

$$csch(C_{CFP}) = L_{LL}$$

$$coth(C_{CFP}) = C_{CFP}$$

$$\frac{L_{LL}\, e^{\sqrt{1+L_{LL}^2}}}{1 + \sqrt{1 + L_{LL}^2}} = 1$$

L_{LL} = the Laplace limit

Archimedes' constant

π = the ratio of a circle's circumference to its diameter.

$$\pi = \frac{\textcolor{red}{red}}{\textcolor{blue}{blue}}$$

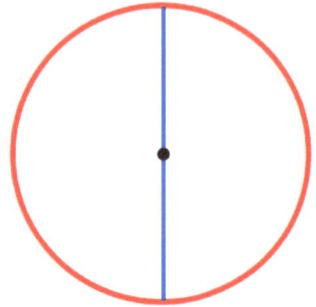

$$\infty = \frac{2}{1}\,\frac{4}{3}\,\frac{6}{5}\,\frac{8}{7}\,\frac{10}{9}\,\frac{12}{11}\,\frac{14}{13}\,\frac{16}{15}\,\frac{18}{17}\,\frac{20}{19}\,\frac{22}{21}\,\frac{24}{23}\,\frac{26}{25}\,\frac{28}{27}\,\frac{30}{29}\,\frac{32}{31}\cdots$$

$$0 = \frac{2}{3}\,\frac{4}{5}\,\frac{6}{7}\,\frac{8}{9}\,\frac{10}{11}\,\frac{12}{13}\,\frac{14}{15}\,\frac{16}{17}\,\frac{18}{19}\,\frac{20}{21}\,\frac{22}{23}\,\frac{24}{25}\,\frac{26}{27}\,\frac{28}{29}\,\frac{30}{31}\,\frac{32}{33}\cdots$$

$$\frac{\pi}{2} = \left(\frac{2}{1}\,\frac{2}{3}\right)\left(\frac{4}{3}\,\frac{4}{5}\right)\left(\frac{6}{5}\,\frac{6}{7}\right)\left(\frac{8}{7}\,\frac{8}{9}\right)\left(\frac{10}{9}\,\frac{10}{11}\right)\left(\frac{12}{11}\,\frac{12}{13}\right)\cdots$$

$$\frac{\pi}{4} = \frac{1}{1} - \frac{1}{3} + \frac{1}{5} - \frac{1}{7} + \frac{1}{9} - \frac{1}{11} + \frac{1}{13} - \frac{1}{15} + \frac{1}{17} - \frac{1}{19} + \frac{1}{21} - \frac{1}{23} + \cdots$$

$$\frac{\pi^2}{6} = \frac{1}{1^2} + \frac{1}{3^2} + \frac{1}{5^2} + \frac{1}{7^2} + \frac{1}{9^2} + \frac{1}{11^2} + \frac{1}{13^2} + \frac{1}{15^2} + \frac{1}{17^2} + \frac{1}{19^2} + \cdots$$

$$\pi = \int_{-1}^{1} \frac{1}{\sqrt{1-x^2}}\,dx \qquad\qquad \frac{\pi}{8} = \int_{0}^{1} \sqrt{x(1-x)}\,dx$$

$$\pi = \int_{-\infty}^{\infty} \frac{1}{1+x^2}\,dx \qquad\qquad \frac{\pi}{2} = \int_{-1}^{1} \sqrt{1-x^2}\,dx$$

$$\pi = \log(e^\pi) \qquad\qquad \sqrt{\pi} = \int_{-\infty}^{\infty} e^{-x^2}\,dx$$

$$2\pi = ж_1 ж_2 ж_3 ж_4 \qquad\qquad \sqrt{2\pi} = e^{-\zeta'(0)}$$

$$-4\pi = ж_1{}^2 + ж_2{}^2 + ж_3{}^2 + ж_4{}^2 \qquad\qquad 0 = e^{\pi i} + 1$$

$$\frac{2}{\pi} = \frac{\sqrt{2}}{2} \cdot \frac{\sqrt{2+\sqrt{2}}}{2} \cdot \frac{\sqrt{2+\sqrt{2+\sqrt{2}}}}{2} \cdot \cdots \qquad \pi = 3 + \cfrac{1^2}{6 + \cfrac{3^2}{6 + \cfrac{5^2}{6 + \cfrac{7^2}{6 + \cfrac{9^2}{6 + \cdots}}}}}$$

Where ∞ = infinity, π = Archimedes' constant, $ж_1$, $ж_2$, $ж_1$, and $ж_4$ = the 1st, 2nd, 3rd, and 4th hyperbolic vortex partition constants, e = Euler's number, $\log(x)$ = the hyperbolic logarithm function, and $\zeta(s)$ = the Reimann zeta function.

the universal parabolic constant

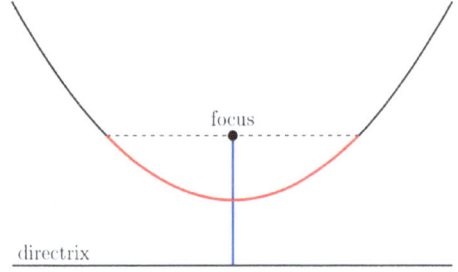

$$P_{up} = \frac{\textcolor{red}{red}}{\textcolor{blue}{blue}}$$

$$\frac{P_{up}}{6} = 8 \int_0^{\frac{1}{2}} \int_0^x \sqrt{x^2 + y^2} \; dy \, dx$$

$$P_{up} = \int_{-1}^1 \sqrt{1 + x^2} \, dx$$

$$\bar{S}_{lse} = \frac{2}{3} \int_0^1 \int_0^1 \sqrt{x^2 + y^2} \; dx \, dy + \frac{1}{3} \int_0^1 \int_0^1 \sqrt{1 + (y - z)^2} \; dz \, dy$$

$$-\bar{S}_{lse} = -\frac{1}{3}\left(\frac{1}{3} P_{up} + \frac{2}{3}\left(1 + \left(\frac{P_{up} - \sqrt{2}}{2} \right) \right) \right)$$

$$P_{up} = \sqrt{2} + \log(\sqrt{2} + 1) \qquad\qquad \text{universal parabolic constant}$$

$$\log(\sqrt{2} + 1) = sinh^{-1}(1) = cosh^{-1}(2)$$

Where $\frac{P_{up}}{6}$ defines the average distance from a point randomly selected in a unit square to its center, $\log(x) =$ the hyperbolic logarithm function, $sinh(x) =$ the hyperbolic sine function, and $cosh(x) =$ the hyperbolic cosine function.

164

the golden ratio

$$\varphi = \sqrt{1 + \sqrt{1 + \sqrt{1 + \sqrt{1 + \sqrt{1 + \sqrt{1 + \sqrt{1 + \sqrt{1 + \sqrt{1 + \sqrt{1 + \cdots}}}}}}}}}}$$

$$\varphi = 1 + \cfrac{1}{1 + \cfrac{1}{1 + \cfrac{1}{1 + \cfrac{1}{1 + \cfrac{1}{1 + \cfrac{1}{1 + \cfrac{1}{1 + \cfrac{1}{1 + \cfrac{1}{1 + \cfrac{1}{1 + \cfrac{1}{1 + \cfrac{1}{1+\ddots}}}}}}}}}}}}$$

$$\varphi + 1 = \varphi^2$$

$$\varphi - 1 = \frac{1}{\varphi}$$

$$sinh(\log(\varphi)) = \frac{1}{2}$$

$$\frac{1}{\varphi} = 0 + \cfrac{1}{1 + \cfrac{1}{1 + \cfrac{1}{1 + \cfrac{1}{1 + \cfrac{1}{1 + \cfrac{1}{1 + \cfrac{1}{1 + \cfrac{1}{1 + \cfrac{1}{1 + \cfrac{1}{1 + \cfrac{1}{1 + \cfrac{1}{1+\ddots}}}}}}}}}}}}$$

$$\varphi = 2\cos\left(\frac{\pi}{n}\right)$$

$$\varphi = \frac{1}{2}\csc\left(\frac{\pi}{2n}\right)$$

$$\varphi = \frac{1}{2}\sec\left(\frac{2\pi}{n}\right)$$

$$\varphi = \frac{1 + \sqrt{n}}{2}$$

the imaginary unit

$$i^2 = -1 \qquad\qquad e^\pi = (-1)^{-i} \qquad\qquad \left(\left(i^{i^2}\right)^2\right)^2 = 1$$

$$i^i = e^{-\frac{\pi}{2}} \qquad\qquad i^{i^2} = -i \qquad\qquad \left(\left(i^i\right)^2\right)^2 = e^{-2\pi}$$

$$i^{-i} = e^{\frac{\pi}{2}} \qquad\qquad \left(i^i\right)^2 = e^{-\pi} \qquad\qquad \left(\left(\left(i^i\right)^2\right)^2\right)^2 = e^{-4\pi}$$

$$\left(i^{-i}\right)^2 = e^\pi \qquad\qquad \left(i^{i^2}\right)^2 = -1 \qquad\qquad 0 = e^{\pi i} + 1$$

$$i^{-i} = 2\, L^4\, W_{\text{We}}{}^4$$

$$e^{xi} = \cos(x) + i\sin(x)$$

$$\left|\, i!\,\right| = \left|\,\Gamma(1+i)\,\right| = \sqrt{\frac{\pi}{\sinh(\pi)}}$$

$$\cos(i) = \cosh(1) \qquad\qquad\qquad\qquad \sin(i) = i\sinh(1)$$

$$\coth^{-1}(0) = -\frac{\pi}{2}i \qquad\qquad\qquad\qquad \coth^{-1}(i) = -\frac{\pi}{4}i$$

$$\log_i(x) = \left(\frac{2}{\pi i}\right)\log(x)$$

$$i^{i^{i^{\cdots}}} = -\frac{W(-\log(i))}{\log(i)} = \left(\frac{2}{\pi}i\right)W\left(-\frac{\pi}{2}i\right)$$

Where π = Archimedes' constant, e = Euler's number, L = the lemniscate constant, W_{We} = the Weierstrass constant, $\Gamma(s)$ = the gamma function, $\sinh(x)$ = the hyperbolic sine function, $x!$ = the factorial function, $|x|$ = the absolute value function, $\log(x)$ = the hyperbolic logarithm function, and $W(x)$ = the product log function.

the hyperbolic logarithm

$$\log(ab) = \log(a) + \log(b)$$

$$\log(n!) = \log(1) + \log(2) + \log(3) + \log(4) + \cdots + \log(n)$$

$$\log(1 + x) = \frac{x^1}{1} - \frac{x^2}{2} + \frac{x^3}{3} - \frac{x^4}{4} + \frac{x^5}{5} - \frac{x^6}{6} + \frac{x^7}{7} - \frac{x^8}{8} + \frac{x^9}{9} - \cdots$$

$$\log(x) = \frac{x^{-1}}{1} - \frac{x^{-2}}{2} + \frac{x^{-3}}{3} - \frac{x^{-4}}{4} + \frac{x^{-5}}{5} - \frac{x^{-6}}{6} + \frac{x^{-7}}{7} - \frac{x^{-8}}{8} + \cdots$$

$$\log(n) = \int_1^n \frac{1}{x}\, dx \qquad\qquad 1 = \int_0^e \frac{1}{x}\, dx$$

$$\log(\cos(x) + i\sin(x)) = i\,x \qquad 1 = x\cosh(\log x) - x\sinh(\log x)$$

$$\frac{1}{x} + x = e^{\log x} + e^{-\log x} = 2\cosh(\log x)$$

$$\log_i(x) = \frac{2\log(x)}{\pi\, i} \qquad\qquad \log_i(e^\pi) = -2i$$

$$Ei(\log x) = li(x) \qquad\qquad li(e^x) = Ei(x)$$

$$\sinh(\log(\varphi)) = \frac{1}{2} \qquad\qquad \log(-1) = \pi i$$

$$\log(2)^2 = Li_1(-1)^2$$

$$\log(2) = Li_1\left(\frac{1}{2}\right) = \sum_{k=1}^{\infty} \frac{(-1)^{k-1}}{k} = \sum_{k=1}^{\infty} \frac{(-1)^{k+1}}{k}$$

$$\log(2) = 2n\left(\sum_{k=1}^{\infty} \frac{1}{k(e^{\pi k} + 1)} + \frac{3}{n} \sum_{k=1}^{\infty} \frac{1}{k(e^{\pi k} - 1)} - \frac{2}{n} \sum_{k=1}^{\infty} \frac{1}{k(e^{2\pi k} + 1)} \right)$$

the polylogarithm

$$Li_s(x) = \frac{x^1}{1^s} + \frac{x^2}{2^s} + \frac{x^3}{3^s} + \frac{x^4}{4^s} + \frac{x^5}{5^s} + \frac{x^6}{6^s} + \frac{x^7}{7^s} + \frac{x^8}{8^s} + \frac{x^9}{9^s} + \cdots$$

$Li_s(1) = \zeta(s)$ $\qquad\qquad\qquad\qquad$ $\zeta(s)$ = the Reimann zeta function

$Li_s(-1) = -\eta(s)$ $\qquad\qquad\qquad\qquad$ $\eta(s)$ = the Dirichlet eta function

$Li_1(x) = -\log(1-x)$ $\qquad\quad$ $\log(x)$ = the hyperbolic logarithm function

$$Li_0(x) = \frac{x}{1-x}$$

$$Li_{-1}(x) = \frac{x}{(1-x)^2}$$

$$Li_{-2}(x) = \frac{x(1+x)}{(1-x)^3} \qquad\qquad\qquad Li_2(0) = 0$$

$$Li_1\left(\frac{1}{2}\right) = \log(2) \qquad\qquad\qquad Li_2(-1) = -\frac{\pi^2}{12}$$

$$Li_2\left((-1)^{\frac{1}{3}}\right) = \left(\frac{4\pi}{\Gamma(n)}\right)^2 + G_{Gi}\, i$$

$$V_{fe} = i\left[-Li_2\left((-1)^{\frac{1}{3}}\right) + Li_2\left(-(-1)^{\frac{2}{3}}\right)\right]$$

$$Li_5\left(\frac{1}{2}\right) = -\zeta(-1,-1,1,1,1) \quad \zeta(s_1, \dots s_k) = \text{the multiple zeta function}$$

$$\int_{1-\varphi}^{-\varphi} \frac{Li_2(x)}{x}\,dx = Li_3(-\varphi) - Li_3(1-\varphi) \qquad \text{golden dilogarithm integral}$$

$$\int_{-\frac{1}{\varphi}}^{-\varphi} \frac{Li_2(x)}{x}\,dx = Li_3(-\varphi) - Li_3\left(-\frac{1}{\varphi}\right) \qquad \text{golden dilogarithm integral}$$

$$Im\left(Li_s(x)\right) = -\frac{\pi(\log(x))^{s-1}}{\Gamma(s)}$$

the Silverman constant

$$S_S = \sum_{n=1}^{\infty} \frac{1}{\phi(n)\,\sigma_1(n)} = \prod_{p \text{ prime}} \left(1 + \sum_{k=0}^{\infty} \frac{1}{p^{2k} - p^{k-1}}\right)$$

$$\phi(\sigma_1(n)) = n \qquad\qquad\qquad \text{the totient of the divisor}$$

$$\sigma_k(n) = \sum_{d|n} d^k \qquad\qquad\qquad \text{the divisor function}$$

$$\sum_{n=1}^{\infty} \frac{\sigma_0(n)}{n^s} = \zeta(s)^2 \qquad\qquad\qquad \text{for } s > 1$$

$$\sum_{n=1}^{\infty} \frac{\sigma_1(n)}{n^s} = \zeta(s)\,\zeta(s-1) \qquad\qquad\qquad \text{for } s > 2$$

$$\sum_{n=1}^{\infty} \frac{\sigma_k(n)}{n^s} = \zeta(s)\,\zeta(s-k) \qquad\qquad\qquad \text{for } s > 2, \text{and } k \geq 0$$

$$\liminf_{n \to \infty} \phi(n)\,\frac{\log(\log(n))}{n} = e^{-\gamma}$$

$$\overline{\lim}_{n \to \infty} \frac{\sigma_1(n)}{n \log(\log(n))} = e^{\gamma}$$

Where $\phi(n)$ = Euler's totient function (defining the number of positive integers $\leq n$ that are relatively prime to n), $\sigma_1(n)$ = the divisor function (defined as the sum of the kth powers of the positive integer divisors of n), p = prime numbers, $\log(x)$ = the hyperbolic logarithm function, $\zeta(s)$ = the Reimann zeta function, e = Euler's number, and γ = the Euler-Mascheroni constant.

Note: The Reimann hypothesis is equivalent to the statement that

$$\frac{\sigma_1(n)}{n \log(\log(n))} < e^{\gamma}$$

decomposition map of the minimal arena

$$(2\cosh(\log b))^{-1}\left(\cosh\left(\frac{n}{2}\right)\right)^{-2}\left(\cos\left(\frac{b}{n}\right)\right)^{-2}e^{-\phi_4}=d_4^{-1}\quad(4^{th})^{-1}$$

$$(\text{external reference})=1$$

$$2\pi\,n\left(\cos\left(\frac{b}{n}\right)\right)^{2}e^{\phi_3}=d_3\qquad\qquad 3^{rd}$$

$$\frac{n}{\sqrt{b\pi}\,(3)^{1/3}}(\,2^{2n}\,e^{\pi}\,)^{-1/8}\left(\Gamma\left(\left(\frac{1}{2}\right)^{2}\right)\right)^{2}e^{\phi_2}=d_2\qquad\qquad 2^{nd}$$

$$2\cosh(\log b)\left(\cosh\left(\frac{n}{2}\right)\right)^{2}\left(\cos\left(\frac{b}{n}\right)\right)^{2}e^{\phi_4}=d_4\qquad\qquad 4^{th}$$

$$\left(\sinh\left(\sinh\left(\frac{1}{b}\right)\right)\right)^{-1}e^{\phi_1}=d_1\qquad\qquad 1^{st}$$

$$\pi\left(\sinh\left(\left(\frac{1}{2}\right)^{2}\right)\right)^{2}e^{\phi_0}=d_0\qquad\qquad 0^{th}$$

Where ϕ_k = the external rotation of the kth balance, d_k = the number of derangements participating in the kth balance, $n=5$ the number of unique rotations maintained by the partition balance of the minimal arena, $!n=44$ the number of derangements available to 5 rotations, $b=7$ the break in scale symmetry between the internal 2 balances of that arena, π = Archimedes' constant, e = Euler's number, $\sinh(x)$ = the hyperbolic sine function, $\cosh(x)$ = the hyperbolic cosine function, $\cos(x)$ = the cosine function, $\log(x)$ = the hyperbolic/natural logarithm function, and $\Gamma(s)$ = the gamma function, which encodes hyperbolically balanced partitions.

$\phi_4 = 1.4167869859079\ldots + (2\pi k)i$ $\quad d_4 = 2(\sqrt{d_0-d_1}+1)^2 = 32$

$\phi_3 = 2.1764268381757\ldots + (2\pi k)i$ $\quad d_3 = 2(\sqrt{d_0-d_1}-1)^2 = 8$

$\phi_2 = 1.8755459671396\ldots + (2\pi k)i$ $\quad d_2 = 2(\sqrt{d_0-d_1}\pm 0)^2 = 18$

$\phi_1 = 1.6162591817564\ldots + (2\pi k)i$ $\quad d_1 = bn = 35$

$\phi_0 = 5.3912583683231\ldots + (2\pi k)i$ $\quad d_0 = !n = 44$

the partition parameters of the minimal arena

$n = 5$ — number of unique rotations
$b = 7$ — break in scale symmetry

$d_0 = {!}\,n = 44$ — (full derangement) derangements of balance 0
$d_1 = bn = 35$ — derangements of balance 1
$d_2 = 2(\sqrt{d_0 - d_1} \pm 0)^2 = 18$ — derangements of balance 2
$d_3 = 2(\sqrt{d_0 - d_1} - 1)^2 = 8$ — derangements of balance 3
$d_4 = 2(\sqrt{d_0 - d_1} + 1)^2 = 32$ — derangements of boundary 4

$\phi_0 = 5.39125836832313\ldots + (2\pi k)i \quad k \in \mathbb{Z}$ — 0th external rotation
$\phi_1 = 1.61625918175645\ldots + (2\pi k)i$ — 1st external rotation
$\phi_2 = 1.87554596713962\ldots + (2\pi k)i$ — 2nd external rotation
$\phi_3 = 2.17642683817579\ldots + (2\pi k)i$ — 3rd external rotation
$\phi_4 = 1.41678698590795\ldots + (2\pi k)i$ — 4th external rotation

$t_P = 5.39125836832313\ldots \times 10^{-44}\ s$ — Planck time
$l_P = 1.61625918175645\ldots \times 10^{-35}\ m$ — Planck length
$q_P = 1.87554596713962\ldots \times 10^{-18}\ C$ — Planck charge
$m_P = 2.17642683817579\ldots \times 10^{-8}\ kg$ — Planck mass
$T_P = 1.41678698590795\ldots \times 10^{32}\ K$ — Planck temperature

$G_{Gi} = 1.01494160640965\ldots$ — Gieseking's constant
$V_{fe} = 2.02988321281930\ldots$ — figure eight knot complement volume
$e = 2.71828182845904\ldots$ — Euler's number
$\pi = 3.14159265358979\ldots$ — Archimedes' constant
$P_{up} = 2.29558714939263\ldots$ — universal parabolic constant
$L = 2.622057554292119\ldots$ — lemniscate constant
$Ж_1 = 0.0854245431533304\ldots$ — 1st hyperbolic vortex partition constant
$Ж_2 = 3.66756753485499\ldots$ — 2nd hyperbolic vortex partition constant
$Ж_3 = -1.87649603900417\ldots + 4.06615262615972\ldots i$ — 3rd hvpc
$Ж_4 = -1.87649603900417\ldots - 4.06615262615972\ldots i$ — 4th hvpc
$Ж_r = 4.47826244916751\ldots$ — hyperbolic vortex radius constant
$Ж_\theta = 2.00316562310924\ldots$ — hyperbolic vortex radian constant
$\varphi = 1.61803398874989\ldots$ — the golden ratio
$\gamma = 0.577215664901532\ldots$ — Euler-Mascheroni constant
$\bar{s}_{lse} = 0.869009055274534\ldots$ — mean line-between-square-edges length
$\alpha_F = 2.50290787509589\ldots$ — alpha Feigenbaum constant
$\delta_F = 4.66920160910299\ldots$ — delta Feigenbaum constant
$W_{We} = 0.474949379987920\ldots$ — Weierstrass constant
$\mu = 1.45136923488338\ldots$ — nontrivial zero of the logarithmic integral
$C_{CFP} = 1.19967864025773\ldots$ — Real fixed point of the hyperbolic cotangent

172

$L_{LL} = 0.662743419349181\ldots$ — Laplace limit
$A_h{}^* = 7.25694640486057\ldots$ — dimension of maximal n-hypersphere area
$V_h{}^* = 5.25694640486057\ldots$ — dimension of maximal n-hypersphere volume
$S_S = 1.78657645936592\ldots$ — Silverman constant
$D_{DHA} = 0.807945506599034\ldots$ — offset of 2 unit disks overlapping by half
$G_g = 1.15872847301812\ldots$ — tether length for grazing half the unit circle
$i^{i^{i^{\cdots}}} = 0.438282936727032\ldots + 0.360592471871385\ldots i$ — i power tower
$C_{R1} = 0.998136044598509\ldots$ — Ramanujan's 1st continued fraction constant
$\Gamma(x_{min}) = 0.885603194410888\ldots$ — minimal value of Γ for $+$ argument
$x_{min} = 1.461632144968362\ldots$ — value at which Γ is minimal for $+$ argument
$F_{FR} = 2.80777024202851\ldots$ — Fransén-Robinson constant
$m_R = 0.7475979202534114\ldots$ — Rényi's parking constant
$\omega_1 = 0.764977018528596\ldots + 1.32497062714087\ldots i$ — omega_1 constant
$\omega_2 = 1.529954037057192\ldots$ — omega_2 constant
$W(1) = 0.567143290409783\ldots$ — the omega constant
$C_{PTA} = 0.531339949958421\ldots$ — Pythagorean triple constant for areas
$\rho_1 = 0.5 + 14.1314251417346\ldots i$ — 1st nontrivial zero of the zeta function
$\zeta(2) = \pi^2/6$ — zeta of 2
$\zeta(3) = 1.20205690315959\ldots$ — Aprey's constant
$\lambda_{GD} = 0.624329988543550\ldots$ — Golomb-Dickman constant
$F_{FF} = 1.22674201072035\ldots$ — Fibonacci factorial constant
$C_{Murata} = 2.82641999706759\ldots$ — Murata's constant

17 mass partitions

$m_e = 9.10938370161994\ldots \times 10^{-31}\ kg$ — electron mass
$m_+ = 1.67262192371195\ldots \times 10^{-27}\ kg$ — proton mass
$m_N = 1.67492749802284\ldots \times 10^{-27}\ kg$ — neutron mass
$m_\mu = 1.88353163790140\ldots \times 10^{-28}\ kg$ — muon mass
$m_Z = 1.62556312860926\ldots \times 10^{-25}\ kg$ — Z boson mass
$m_W = 1.43310146854387\ldots \times 10^{-25}\ kg$ — W boson mass
$m_\tau = 3.16754377503113\ldots \times 10^{-27}\ kg$ — tau mass
$m_H = 2.23149658442420\ldots \times 10^{-25}\ kg$ — Higgs boson mass
$m_t = 3.08638372655297\ldots \times 10^{-25}\ kg$ — truth (top) quark mass
$m_c = 2.28313100509581\ldots \times 10^{-27}\ kg$ — charm quark mass
$m_b = 7.48705738029755\ldots \times 10^{-27}\ kg$ — beauty (bottom) quark mass
$m_s = 1.88130136459184\ldots \times 10^{-28}\ kg$ — strange quark mass
$m_u = 4.00077202715006\ldots \times 10^{-30}\ kg$ — up quark mass
$m_d = 8.61183215072197\ldots \times 10^{-30}\ kg$ — down quark mass
$m_{\nu_\tau} = ???\ldots \times 10^{-??}\ kg$ — tau neutrino mass
$m_{\nu_\mu} = ???\ldots \times 10^{-??}\ kg$ — muon neutrino mass
$m_{\nu_e} = ???\ldots \times 10^{-??}\ kg$ — electron neutrino mass

constants of Nature

$\alpha = 7.29735257295522 \ldots \times 10^{-3}$	fine-structure constant
$e = 1.60217657405973 \ldots \times 10^{-19} \, C$	electron charge
$\alpha_G = 1.75182147492404 \ldots \times 10^{-45}$	gravitational coupling constant
$\mu_0 = 1.25663706140197 \ldots \times 10^{-6} \, m \, kg/C^2$	magnetic constant
$S_{mi} = 4.41899541452692 \ldots \times 10^{9} \, kg/s \, C$	Schwinger magnetic induction
$R_K = 2.5812807449400712 \ldots \times 10^{4} \, m^2 kg/s \, C^2$	von Klitzing constant
$H_C = 3.87404586641820 \ldots \times 10^{-5} \, sC^2/m^2 kg$	quantized Hall conductance
$\Phi_0 = 2.06783384793306 \ldots \times 10^{-15} \, m^2 kg/s \, C$	magnetic flux constant
$K_J = 4.8359784854056 \ldots \times 10^{14} \, s \, C/m^2 kg$	Josephson constant
$q_c = 3.63694755140949 \ldots \times 10^{-4} \, m^2/s$	quantum of circulation
$G_0 = 7.74809172907834 \ldots \times 10^{-5} \, s \, C^2/m^2 kg$	conductance quantum
$c_2 = 1.43877687654906 \ldots \times 10^{-2} \, m \, K$	2nd radiation constant
$\hbar = 1.05457172603011 \ldots \times 10^{-34} \, m^2 kg/s$	Planck's constant
$\kappa = 8.98755179228829 \ldots \times 10^{9} \, m^3 kg/s^2 C^2$	Coulomb's constant
$\varepsilon_0 = 8.85418781277322 \ldots \times 10^{-12} \, s^2 C^2/m^3 kg$	electric constant
$c = 2.99792458354727 \ldots \times 10^{8} \, m/s$	speed of light
$F = 9.64853321242908 \ldots \times 10^{4} \, C/mol$	Faraday constant
$g_N = -3.82608560204984 \ldots$	neutron g-factor
$G = 6.67384038951738 \ldots \times 10^{-11} \, m^3/s^2 kg$	gravitational constant
$k_B = 1.38064931695409 \ldots \times 10^{-23} \, m^2 kg/s^2 K$	Boltzmann constant
$\omega_c = 7.7634409944556 \ldots \times 10^{20} \, 1/s$	Compton angular frequency
$c_{1L} = 1.19104287769811 \ldots \times 10^{-16} \, m^4 kg/s^3$	spectral radiance
$c_1 = 3.74177185197781 \ldots \times 10^{-16} \, m^4 kg/s^3$	1st radiation constant
$Z_0 = 3.76730313666885 \ldots \times 10^{2} \, m^2 kg/s \, C^2$	characteristic impedance
$a_0 = 5.29177210889649 \ldots \times 10^{-11} \, m$	Bohr electron radius
$r_e = 2.81794032505462 \ldots \times 10^{-15} \, m$	classical electron radius
$E_h = 4.359744722203833 \ldots \times 10^{-18} \, m^2 kg/s^2$	Hartree energy
$\lambda_C = 2.42631023893941 \ldots \times 10^{-12} \, m$	Compton wavelength
$\mu_B = 9.27400999397153 \ldots \times 10^{-24} \, m^2 C/s$	Bohr magneton
$\mu_N = 5.05078369897026 \ldots \times 10^{-27} \, m^2 C/s$	Nuclear magneton
$N_A = 6.02214076693260 \ldots \times 10^{23} \, 1/mol$	Avogadro constant
$\sigma = 5.67037441935166 \ldots \times 10^{-8} \, kg/s^3 K^4$	Stefan-Boltzmann constant
$R = 8.31446261914764 \ldots \, m^2 kg/s^2 K \, mol$	molar gas constant
$\gamma_+ = 2.67522187383442 \ldots \times 10^{8} \, s/C \, kg$	proton gyromagnetic ratio
$m_u = 1.66053906637533 \ldots \times 10^{-27} \, kg$	atomic mass constant
$\sigma_e = 6.65246160010799 \ldots \times 10^{-29} \, m^2$	electron Thomson x section
$R_\infty = 1.09737315685250 \ldots \times 10^{7} \, 1/m$	Rydberg constant
$g_\mu = -2.00233184124323 \ldots$	muon g-factor
$g_e = -2.00231930436319 \ldots$	electron g-factor
$g_+ = 5.58569468885333 \ldots$	proton g-factor
$N_\mu = -9.66236470868258 \ldots \times 10^{-27} \, m^2 C/s$	neutron magnetic moment
$b_{entropy} = 3.002916077148106 \ldots \times 10^{-3} \, mK$	Wien entropy constant
$r_N = 7.72554339837953 \ldots \times 10^{-16} \, m$	neutron radius

$r_+ = 8.40702954466144 \ldots \times 10^{-16}\, m$ proton radius

Where the black digits represent previously known values (either measured or geometrically known), and green digits represent extended predictions.

total number of predicted digits $= 44 \times 14 = 616$
number of measurement digits needing explanation $= 442$
number of predicted digits matching measurement $= 442$
number of additionally predicted digits $= 174$

Other books by Thad:

Einstein's Intuition: Visualizing Nature in Eleven Dimensions

Moon Rock: Mare Crisium

Passages

A Perfect Universe

Source Code: the balance of persistence